551.46
T11o

81216

DATE DUE			
Nov 1 '73	Aug 10 '82		
Dec 11 '73			
Mar 28 '74			
Jul 29 '74			
Jan 13 '75			
Feb 4 '76			
Nov 15 '76			
9/6/77			
Oct 26 79			
Nov 14 79			
Nov. 26 '79			
Dec 14 79			

Books in 1001 Questions Answered Series

1001 Questions Answered About the Oceans and Oceanography

1001 Questions Answered About the Oceans and Oceanography

BY ROBERT W. TABER AND HAROLD W. DUBACH

Foreword by Richard H. Fleming

ILLUSTRATED WITH PHOTOGRAPHS AND DRAWINGS

DODD, MEAD & COMPANY
NEW YORK

ISBN: 0-396-06496-5

LIBRARY OF CONGRESS CATALOG CARD NUMBER: 73-184136

PRINTED IN THE UNITED STATES OF AMERICA

BY VAIL-BALLOU PRESS, INC., BINGHAMTON, N.Y.

—to the memory of Ralph Wheaton Whipple, Professor of Natural Science at Marietta College from 1919 to 1954.

—to Hubert Campbell, former instructor, Lafayette High School, St. Joseph, Missouri.

FOREWORD

Questions are the basic tools of learning. This is true for the small child with his endless queries, for student and teacher in the classroom, and for the scientist whose lifelong quest for truth and understanding is guided by the questions that he poses to himself.

Asking questions is easy, but it is not so simple to ask good questions. What is a good question? It is one that has been carefully thought out and worded in such a way that, when answered, it stimulates further questions and answers. When poorly worded a question may evoke an inappropriate reply and hence mislead its asker. Learning is a two-way process. The best questions are those that encourage this exchange so that both questioner and respondent gain in wisdom and perception.

In matters concerning the ocean—and in fact all aspects of nature—it is helpful to identify several different categories of knowledge. As a mnemonic device these can be called *what, where, when, who,* and *why. What* refers to all properties and features. *Where* includes all the topics that deal with the location or distribution of features and properties with reference to the earth's coordinate system. The *when* category relates an event to the scale of time or emphasizes the importance of the manner in which features and properties (and their distributions) change with time. *Who* concerns the roles of individuals, ships, institutions, and their relation to their environments.

To answer questions that involve *what, where, when,* and *who,* the information needed is descriptive, that is, the kind of knowledge that is accumulated from systematic observations and measurements and has then been compiled on maps and charts, in books, and, in recent times, in computers. If the facts are known, the main difficulty may be to find the specific information to answer a question.

The topics that fall within the *why* category may well pose difficulties to both questioner and answerer. Science is primarily concerned with the search for answers to questions of this kind. The quest for knowledge and understanding of man and his environment is as old as the human race. Much has been learned, many answers found, but as is true of all good questions, new answers raise further questions.

The best advice that I can offer to the reader of this volume is to ask questions, to seek answers, but never to be satisfied with his new understanding. Continue the search for wisdom—ask more and better questions. Soon you will find that you can answer many of them yourself. You are then embarked on the long but enjoyable voyage of the scholar. If you crave adventure and the exploration of the unknown, where you can both ask questions that are new and original and then yourself seek to find the answers, join those who cruise under the flag of science.

Richard H. Fleming
Professor
Department of Oceanography
University of Washington

CONTENTS

PHOTOGRAPHS

(Following page 80)

1001 Questions Answered About the Oceans and Oceanography

I. THE WORLD OCEAN

1. How many oceans are there? Oceanographers divide the world ocean into three oceans, the Atlantic, Pacific, and Indian. There has been a tendency to divide the world ocean into seven oceans to retain the legendary number associated with the seven seas. The popular division is Arctic, North Atlantic, South Atlantic, North Pacific, South Pacific, Indian, and Antarctic. The International Hydrographic Bureau at Monaco does not accept the Antarctic as a separate ocean. Actually, of course, all limits of oceans are arbitrary, as there is only one global sea.

2. When were the oceans named? It was not until 1845 that the names *Atlantic, Pacific,* and *Indian* were finally accepted for the three major oceans. In the past the Atlantic was called the *Great Ocean;* the North Atlantic was called the *North Ocean* and the South Atlantic was called the *South Ocean.* The Pacific was once known as the *Western Ocean.*

3. What and where is the Southern Ocean? This name is often used to refer to the waters surrounding Antarctica; it is also known as the Antarctic Ocean. The Zone of Antarctic Convergence, located at approximately 55° S is considered to be the northern limit. There is, however, no good geographic line of demarcation for the Southern Ocean. Many marine scientists refer to waters bordering the Antarctic Continent as regions of the South Atlantic, the South Pacific, or the Indian Oceans.

4. What is the difference between a sea and an ocean? The terms *sea* and *ocean* are often used interchangeably in referring to salt water. However, from a geographic point of view, a sea is a body of water that is substantially smaller than an ocean or is part of an ocean. The term *seven seas* dates back to ancient times, referring to the seas known to the Mohammedans before the fifteenth century. These were the Mediterranean Sea, the Red Sea, the East African Sea, the West African Sea, the China Sea, the Persian Gulf, and the Indian Ocean. In more recent times, Rudyard Kipling popularized

1

the expression *seven seas* by using it as the title of a volume of poems.

5. How many seas are there?   The International Hydrographic Bureau lists 54 seas; some are seas within seas. The Mediterranean Sea contains seven seas, so one could sail the seven seas (of the Mediterranean) without ever venturing into an ocean. The International Hydrographic Bureau subdivisions are primarily for the purpose of filing *Notices to Mariners* and have little to do with natural boundaries.

6. How did seas such as the Black, Red, and White get their names?   Because the Black Sea is landlocked, it is deficient in oxygen, except near the surface. The high concentration of hydrogen sulfide colors the mud black on the sea floor. Oddly enough, the recurring bloom of small blue-green algae (*Trichodesmium erytraeum*) imparts the red color to the Red Sea. The White Sea received its name from the ice that covers it more than two hundred days a year. The color of the Yellow Sea is caused by the yellow mud which is carried by rivers, especially when floods occur.

7. What is the volume of the world's oceans?   Estimates vary from 317 to 330 million cubic miles; the most reliable sources place the volume at approximately 328 million cubic miles. The volume of all land above sea level is only one eighteenth of the volume of the oceans. If the solid earth were perfectly smooth, the ocean would cover it to a depth of 12,000 feet.

8. How much of the world's water is in the oceans?   According to the Department of the Interior, about 97 percent of the world's water is in the oceans. Most of the remainder is frozen on Antarctica and Greenland. Less than 1 percent of the world supply is directly available for fresh-water use.

9. How much of the earth's surface is covered by oceans?   The oceans cover about 71 percent of the surface of the earth. They cover about 61 percent of the Northern Hemisphere and 81 percent of the Southern Hemisphere. The Pacific Ocean covers almost as much area as the Atlantic Ocean and Indian Ocean combined.

10. What is the greatest depth of the ocean? In 1959 the Soviet vessel *Vityaz* reported a depth of 36,198 feet in the Marianas Trench off Guam. Previous soundings in the area were made by the British survey ship *Challenger* in 1952 (35,640 feet) and the Japanese survey ship *Manshu* in 1927 (32,190 feet). On January 23, 1960, the bathyscaph *Trieste* descended into the Marianas Trench to a depth of 35,800 feet.

11. What is the greatest depth in the Southern Hemisphere? In 1952 the U.S. research vessel *Horizon* recorded a depth of 34,884 feet in the Tonga Trench, south of Samoa Islands.

12. Which is the deepest ocean? The Pacific is the deepest ocean, averaging 4,282 meters (14,048 feet). The Indian Ocean averages 3,963 meters (13,002 feet). The Atlantic is slightly shallower, averaging 3,926 meters (12,880 feet). These figures do not take into account the shallow adjacent seas. The Baltic Sea has a mean depth of only 180.4 feet.

13. Is sea level rising or falling? In most areas it is difficult to determine whether the sea level is rising or the land sinking. However, it is generally believed that the last ice age caused a general lowering of sea level. Studies of the U.S. east coast and South Pacific islands indicate that 16,000 to 18,000 years ago sea level was about 400 feet lower than today. Since then it has been rising; during the past 4,000 years sea level appears to have been rising about 3 inches per century. During the past century the rise has been 4.5, but this may be a short-period trend.

14. How old are the oceans? Sedimentary rocks (deposited under water) have been dated as 3 billion years old. Some scientists suggest that the oceans are almost as old as the earth and that they contained at least part of their water 4 billion or more years ago.

It is now generally believed that the ocean basins received only a small part of their present volume at the time the earth cooled and water vapors condensed; perhaps only 5 to 10 percent. The basins were probably filled gradually as the continents built up, by water coming from the earth's interior as hot springs and volcanic action.

15. What is the *Marginal Sea*?   The *Marginal Sea* is the water area bordering a nation over which it has exclusive jurisdiction except for the right of innocent passage of foreign vessels. The Marginal Sea is referred to by other names such as *Territorial Sea, Adjacent Sea, Marine Belt,* and *Maritime Belt* and its limits have in recent years been the subject of much international discussion. The concept that maritime countries should have the right of dominion over the sea near its coasts so as to contribute to its self-defense was first presented in 1702 by Cornelius Van Bynkershoek, a Dutch jurist. The cannon-shot rule developed as the first practical means of determining the extent of the Marginal Sea. In 1702 the range of a cannon was approximately a marine league or 3 nautical miles; therefore, the term *3-mile limit* is often used in reference to the Marginal Sea. Today, many countries no longer subscribe to the 3-mile limit as the extent of their territorial jurisdiction.

16. What is the difference between *open coast, open port, open sea,* and *open water*?   *Open coast* refers to a portion of the coast that is not sheltered or protected from the sea.

Open port is a port which is not icebound during the winter season.

Open sea refers to that part of the ocean not enclosed by headlands or within a narrow strait. It is that part of the ocean which is outside the territorial jurisdiction of any country.

Open water is a relatively large area of free water in an ice-filled region, more specifically the water will have less than one tenth of the surface covered with floating ice.

17. What does the Latin phrase "Mare Liberum" mean? Literally, it means "the sea free" or "the sea open." *Mare Liberum* is the title of a work written by Hugo Grotius in 1608 in which he contends that the sea is not capable of private dominion. He urged the Roman doctrine of freedom of the high seas. These have been defined and referred to as "the four freedoms" which are: the freedom of navigation, the freedom of fishing, the freedom to lay cables and pipelines, and the freedom to fly over the high seas.

18. Where does the expression "Right of innocent passage" originate and what does it mean? The right of innocent passage is a principle that has been recognized for a long time as an integral part of the law of the sea. The rules pertaining to innocent passage through the territorial sea were adopted during the First Geneva Conference on the Law of the Sea (February 24–April 27, 1958). It assures a foreign vessel the right of navigation through the territorial sea for the purpose of traversing that sea without entering internal waters so long as the passage is not prejudicial to peace, good order, or security of the coastal state. The right of innocent passage also extends to straits used for international navigation that connect two parts of the high seas or the high seas with the territorial waters of another state.

19. Where is the American Mediterranean? Mariners and oceanographers have given this name to the Caribbean Sea and Gulf of Mexico region. It is called *mediterranean* because it is separated from the North Atlantic Ocean by Cuba, Hispaniola, Puerto Rico, and the Lesser Antilles island chain and associated undersea ridges and is enclosed by the shores of the North and South American continents.

20. What is an estuary? A drowned river mouth having a free connection with the open ocean. As fresh water flows from the river into the estuary, salt water from the ocean will intrude under the fresh water as a wedge. The subsurface seawater mixes with the less saline surface water and is carried seaward.

21. What is a bay? The Geneva Convention defines a bay as a well-marked indentation whose penetration is in such proportion to the width of its mouth as to contain landlocked waters and constitute more than a mere curvature of the coast. The area of such an indentation must be as large as, or larger than the semicircle whose diameter is a line drawn across the mouth of the indentation.

22. Where exactly is the coastline? The coastline (or shoreline) according to the National Ocean Survey is the line of contact between the land and the sea. On U.S. Coast and Geodetic Survey (now National Ocean Survey) charts, it is approximately the mean high-water line.

23. What is the difference between a shore and a coast? The term *shore* is applied to water, while *coast* applies primarily to land. *Shore* is the zone from mean low tide to the inner edge of wave-transported sand. *Coast* is the broad zone landward from the shore, including sea cliffs and elevated terraces as well as the lowlands inside the shore.

24. Why is water so important to man? The origin of life on the earth is believed to be the sea; after millions of years of evolution, the tissues of modern man are bathed in a saline solution which closely resembles seawater. Every metabolic and organic bodily process utilizes and requires watery solutions to accomplish necessary functional processes. In addition to being a necessary ingredient for the effective operation of the body, water supplies needed nutritive substances in solution that are essential to plant life and growth. The sea itself provides food such as fish and other sea animals and waterfowl. The marshes and shores also sustain various types of edible vegetation. And as an available mode of transportation the water areas of the world provide access to the shores and hinterland alike.

25. When one speaks of the natural resources of the ocean, just what does this include? The term *natural resources* has been increasingly used by conservationist, environmentalist, political leaders, students, and the public in recent years. Because the term is imprecise, when Public Law 31, 83rd Congress, was placed on the statutes on May 22, 1953, agreement on its meaning was included because it was a legal necessity. As defined in that statute, marine natural resources includes "oil, gas and all other minerals, and fish, shrimp, oysters, clams, crabs, lobsters, sponges, kelp and other marine animal and plant life but does not include water power or the use of water for the production of power." The Convention on the Continental Shelf adopted at Geneva in 1958 defines *natural resources* of the sea as including mineral and other nonliving resources of the seabed and subsoil and living organisms belonging to sedentary species. Specifically this latter definition excludes the so-called bottom fishes and crustacea or other swimming species even though they burrow into the bottom. Only organisms such as coral, sponges, oysters, clams, and other shellfish, or creatures or plants that are constantly in contact with the bottom are included.

26. What is the total coastline of the United States? The place where ocean and land meet is called the coastline. The total coastline of the United States (including Alaska and Hawaii) is 88,633 miles. Of all the states, Alaska has the longest shoreline; it totals 33,904 miles including all the islands. The Hawaii coastline measures 1,052 miles. Of the 48 conterminous states, the Florida coast is the longest, extending 8,426 miles along the Atlantic and Gulf of Mexico.

27. Are there scriptural references to the ocean? There are many Biblical references to the sea, waves, and other natural environmental occurrences, for men of those days were careful observers of events of nature. In some references, the forces of the sea are likened to God's power.

"All the rivers run into the sea; yet the sea is not full; unto the place from whence the rivers come, thither they return again." This verse from Ecclesiastes 1:7 is an Old Testament way of describing the hydrologic cycle. The creation story in the first chapter of Genesis refers to water and oceanic creatures several times. "And God made the firmament and divided the waters which were under the firmament from the waters which were above the firmament: and it was so" (Genesis 1:7). In this part of the creation story, the writer and his contemporaries are fully aware of the moisture content of the atmosphere. Continuing: "And God said, Let the waters under the heaven be gathered together in one place, and let the dry land appear, and it was so" (Genesis 1:9). "And God said, Let the waters bring forth abundantly the moving creature that hath life . . ." (Genesis 1:20). "And God created great whales and every living creature that moveth, which the waters brought forth abundantly, after their kind . . . Be fruitful and multiply and fill the waters in the seas" (Genesis 1:20–21).

Chapters 37 and 38 of the Book of Job raise questions about nature and man's limited knowledge of it. "Or who shut up the sea with doors, when it brake forth" (Job 38:8). "Hast thou entered into the springs of the sea? or hast thou walked in the search of the depth?" (Job 38:16). "Out of whose womb came the ice? . . . The waters are hid as with a stone, and the face of the deep is frozen" (Job 38:29–30). David in his Psalms refers several times to the sea and its nature and character: see Chapters 65, 66, 89, 93, and 104. In Chapter 107, verses 23 and 24, David writes: "They that go down to the sea in ships, that do business in great waters; these see the works of the Lord, and his wonders in the deep." These verses have been often used by recent writers in referring to mariners and oceanographers. These are but a few of many references to the sea, water, and the marine environment that appear in the Bible.

II. SCIENCE OF THE SEA

28. What is the difference between hydrography and oceanography? To explain the difference between hydrography and oceanography, the ocean can be compared to a bucket of water; then hydrography is the study of the bucket and oceanography is the study of the water. Hydrographers are primarily concerned with the problems of navigation. They chart coastlines and bottom topography. A hydrographic survey usually includes measurements of magnetic declination and dip, tides, currents, and meteorological elements.

Oceanography is concerned with the application of all physical and natural sciences of the sea. It includes the disciplines of physics, chemistry, geography, geology, biology, and meteorology. The interrelationship of specialties is one of the main characteristics of oceanography. Oceanographic and hydrographic surveying may be combined on the same ship. Many times the words *oceanography* and *hydrography* are used interchangeably. Recently the term *oceanology* has been used by many scientists instead of *oceanography* to indicate that it is the study of the oceans instead of only the description of the oceans.

29. What are soundings? Soundings are the depths obtained on a hydrographic survey. These are uncorrected when first recorded; in subsequent processing of the field data, corrections for tide and other factors are made so that the number shown on the finished hydrographic chart is precisely located and referred to the proper chart datum. Until the echo sounder was adopted for use on Navy and merchant ships following World War II, the leadline was used for taking soundings. The echo sounder has been a tremendous aid in hydrographic surveying as it provides a continuous sounding record along each ship track. Commercial ships which employ echo sounders as part of their navigational equipment voluntarily provide many echo sounder records to federal charting agencies. Inclusion of these soundings on charts, however, requires careful checking of the navigational fixes and ship track for proper positioning of the sounding data.

30. What is a leadline?   A sounding lead is attached to a line of sash cord or tiller rope and this is used to measure the water depth. The line is marked at foot and fathom (6 feet) intervals. The bottom of the lead is scooped out; tallow is deposited in that hollow for picking up bottom sediments while sounding.

31. What is echo sounding?   Echo sounding is a method for measuring the depth of a water body by determining the time required for a sound wave to travel from the observing vessel to the bottom and return. The velocity of sound which varies according to the temperature and salinity of the water must be known for the water in which the observations are being made. If the sound velocity in the water is precisely determined, then the depth records obtained by an echo sounder will be accurate.

32. What is a wire drag and how is it used?   This apparatus was developed in the early 1900s by the U.S. Coast and Geodetic Survey (now the National Ocean Survey) for surveying rocky and coral areas where normal sounding methods would be insufficient to ensure discovery of all rocks or other existing obstructions rising above a given depth. The drag consists of a horizontal wire, set to maintain a predetermined depth by use of floats attached at regular intervals along the line, which is towed through the water. The wire will catch on any obstruction rising above the depth at which it is set. The device was first used along inshore areas of the Alaskan coast; it has proven equally useful for shallow tropical waters where coral heads are abundant. Areas surveyed by this procedure receive special annotation on charts indicating that safe navigation is assured only to the depths of the wire drag and that the area has not been adequately sounded. These areas are designated by a green tint on charts.

33. When were the first coastal surveys of the United States done?   As early as 1800, bills were introduced in Congress for specific construction and other harbor work in several areas including the Delaware River and Nantucket Harbor but the first real coastal surveys made by the United States Government were ordered in 1806 and covered the coast of North Carolina and part of Louisiana.

34. What area was the first to be surveyed by the Coast and Geodetic Survey? Though a survey of the coast was authorized by Congress in 1807, nothing was done until 1811. Because funds were not released, nothing could be done and Ferdinand Hassler, the newly appointed superintendent, took a position as acting professor of mathematics at the U.S. Military Academy until 1810 and taught for another year at Union College in Schenectady, New York, before the appropriated $25,000 was made available to begin work. The next year was spent obtaining needed instruments and equipment in Europe; then the War of 1812 prolonged his stay till 1815. Finally, the first survey got under way in 1816 and was to cover the bay and harbor of New York. The two base lines established for this survey were in the vicinity of English Creek, near Englewood, New Jersey, and at Gravesend Village, Long Island, New York.

35. When was the first hydrographic survey made of U.S. waters? The earliest hydrographic survey made by an agency of the United States was a survey done of Boston Harbor by the Navy Department in 1817. Presumably this survey was accomplished under the authority of the Coast Survey (later Coast & Geodetic Survey), as it was a component of the Navy Department at that time. The first survey by Coast Survey ships and personnel was made in 1834 and covered Great South Bay, Long Island, New York. Records of both of these surveys are in the National Ocean Survey Archives.

36. When were the first hydrographic surveys conducted by the Coast and Geodetic Survey? Hydrographic surveys were started in late 1834 and early 1835. In March 1834 the Survey of the Coast was transferred from the Treasury Department to the Navy Department. The Superintendent, Ferdinand Hassler, objected strongly to this transfer, and in exactly two years' time the Survey was returned to the Treasury Department. The first ships assigned to survey duties were the schooner *Jersey* commanded by Lieutenant Gedney and the schooner *Experiment* under Lieutenant George Blake. Again the area was New York Harbor, Great South Bay, and the South Shore of Long Island. That survey discovered many ledges and other features dangerous to navigation and also a previously unknown channel which is appropriately known as Gedney Channel and now serves as a main channel entrance to New York Harbor. The brig *Washington,* built in 1837 as a revenue cutter, was used by the Survey during summers until 1840, at which time she was transferred to the Coast Survey.

37. What was the U. S. Exploring Expedition? This survey, from 1838 to 1842, was conducted by Lieutenant Charles Wilkes. The survey ranged from the eastern Atlantic to Antarctica, the coast of both Americas, and deep into the Western and Southwestern Pacific, contributing substantially to hydrographic, meteorological, botanical, and geological knowledge of the regions explored.

38. When did oceanographic surveys begin? The beginning of modern oceanography is usually considered to be December 30, 1872, when HMS *Challenger* made her first oceanographic station on a 3½-year round-the-world cruise. This was the first purely deep-sea oceanographic expedition ever attempted. Analysis of seawater samples collected on this expedition proved for the first time that the various constituents of salts in seawater are virtually in the same proportion everywhere. This is known as Dittmar's principle.

39. What was the purpose of the *Challenger* Expedition? By the mid-nineteenth century, a British naturalist, Edward Forbes (1815–1854), had made significant contributions to the current marine biological knowledge of his day. His particular interest focused on the vertical distribution of marine life and he hypothesized that there is a zone in the deep beyond which no life exists. This led to great controversy among scientists, which proved to be a helpful stimulus in obtaining authorization for the expedition. As a matter of fact, an international rivalry of the same type that occurred in the "space race" developed. This prompted the Royal Society to appoint a committee in 1871 which recommended that the British Government provide funds for an expedition with these objectives:

To investigate the physical characteristics of the deep water of the ocean basins.

To determine the chemical constituents in seawater at all depths.

To analyze the physical and chemical makeup of the sea floor and determine their origin.

To examine and classify the biota at all depths and on the sea bottom.

40. What were some of the accomplishments of the *Challenger* Expedition?

HMS *Challenger* was a steam corvette of 2,306 tons displacement which was refitted with laboratory space, cables, winches, and other equipment for dredging, sounding, taking *in situ* water temperatures and collecting water samples. The ship was under the command of Captain George Nares; C. Wyville Thompson headed the scientific staff of six. The ship departed on December 7, 1872, from Sheerness, England, and returned 3½ years later on May 24, 1876. The cruise covered almost 69,000 nautical miles in which observations were taken from the Atlantic, Pacific, Indian, and Antarctic (Southern) Oceans. To summarize, observations taken included:

492 deep-sea soundings
362 serial water temperature observations
133 bottom dredges
151 trawls
4,417 new species were discovered

The final report required fifteen years to complete and was issued between 1880 and 1895. It comprises fifty volumes with about 29,500 pages and more than 3,000 illustrations. Contributions from seventy-six authors are included; Sir John Murray was the editor. The entire report was reprinted in 1965 and is being offered for sale at approximately $3,850 by Johnson Reprint Inc., Berkeley Square, London, W1, England.

41. What is the meaning of the words "in-situ" which appear frequently in oceanographic observations?

In situ is a Latin term which means literally *in place*. When used in connection with oceanographic observations it indicates that the observation or measurement of the sample was taken while it remained in its original, natural position in the environment. Oceanographers often wish to measure certain properties of a subsurface water sample at its original location in the ocean because it may be altered when brought to the surface.

42. When was the first U.S. oceanographic ship built?

The first ship built in the United States for oceanographic research was the *Albatross,* launched in 1882. She was built for the U.S. Commission of Fish and Fisheries. Expeditions of the *Albatross* were conducted by Alexander Agassiz, son of Louis Agassiz, the famous zoologist and geologist.

43. What were some of the notable early oceanographic expeditions? The U.S. Naval Oceanographic Office publication *American Practical Navigator* lists the following notable expeditions:

American *Albatross,* 1882–1920
Austrian *Pola* in the Mediterranean and Red Seas between 1890 and 1896
Danish *Dana,* which during its voyages of 1920–22 discovered the breeding place of the European eels in the Sargasso Sea
American *Carnegie* in 1927–1929
German *Meteor* in the Atlantic, 1928–1938
British *Discovery II* in the Antarctic between 1930 and 1939
Norwegian vessels *Fram* and *Maude* in the Arctic ice pack, 1893–1896 and 1918–1925

44. What are some of the important oceanographic expeditions after World War II? Swedish *Albatross,* Danish *Galathea,* Second British *Challenger* (built in 1931) and *Discovery II* in the Antarctic.

45. What did Prince Albert I of Monaco contribute to oceanography? Prince Albert used his personal fortune to finance investigations of the Atlantic Ocean and Mediterranean Sea, beginning in 1885. He was both captain and chief scientist of *Hirondelle.* His lifelong interest was marine biology and his special area of investigation was the giant squid. He established the Oceanographic Museum at Monaco.

46. Where is the Oceanographic Museum? The Oceanographic Museum is located in Monaco-Ville. It is situated on the seaward face of the "Rock," a headland that protects the Port of Monaco from storm waves of the Mediterranean. The museum is said to be the oldest and largest of its kind in the world, is world renowned and stands as a symbol of Monaco's long association with the sea and the science of oceanography. It was built in 1910 by Prince Albert I, then ruler of the principality. He had a lifelong enthusiasm and love for the sea and a dedication to learning more about marine life and their interrelationships with their surroundings. Prince Albert was an active oceanographer and paleontologist and personally led many collecting expeditions in the late nineteenth and early twentieth centuries. Specimens he collected were provided for the museum collections and displays. The museum today is known for its research activities; the present staff includes Captain Jacques-Yves Cousteau, who is world famous for his deep-sea work. The museum includes an aquarium featuring tropical fish and is a major tourist attraction. The current entrance fee is about $1.00.

47. When was the first systematic study of a whole ocean? In 1925 the German *Meteor* began the South Atlantic Expedition. In just over two years, Meteor crossed the South Atlantic more than a dozen times, collecting 70,000 depth soundings which revealed the rugged character of the ocean floor. Seawater samples collected by the *Meteor* revealed that the amount of gold in solution was less than expected and could not be recovered profitably. Oceanographic data collected over this broad ocean area by the *Meteor* are still used by scientists.

48. When did international oceanographic surveys begin? Eight European nations formed the International Council for the Exploration of the Seas (ICES) in 1899 for the purpose of conducting a joint exploration of the North Atlantic, North Sea, and Baltic Sea. But it was not until the International Geophysical Year in the late 1950s that international surveys began on a large scale.

49. What was the IIOE? The International Indian Ocean Expedition was begun in 1959 by thirteen nations. Its purpose was to collect biological specimens on a broad scale by standardized means, so that the results of all nations could be compared. Specimens were sent to the sorting centers of the Indian Ocean Biological Center in Cochin, India, and the Smithsonian Institution in Washington, D.C., for distribution to specialists. Participants of the IIOE also collected a broad spectrum of oceanographic and meteorological observations.

50. What were some of the results of the IIOE? Large unexploited populations of tuna, shrimp, lobster, and sardines were discovered.

Among the many oceanographic discoveries was a current in excess of 6 knots along the Somali coast at the time of the southwest monsoon. At this season, upwelling brings to the surface the coldest water found anywhere in the world so close to the equator.

51. What is the largest oceanographic research ship? The Japanese Arctic Research Ship *Fuji* is the largest ship built for oceanographic research, although larger ships have been converted from other uses. *Fuji,* which was launched in March 1965, has a displacement of 8,305 tons (full load). She was designed for breaking ice more than 20 feet thick, and her bow is heavily armored for driving the ship on top of the ice field and crushing it by sheer weight.

From 1957 until 1965 (when *Fuji* was launched) the Russian Oceanographic Research Ship *Mikhail Lomonosov* was the largest ship designed for oceanographic work. That ship has 16 scientific laboratories capable of performing any type of investigation or analysis. The scientific staff of 69 includes women scientists. Displacement is 5,960 tons.

The largest U.S. oceanographic ships are *Discoverer* and *Oceanographer* of the National Ocean Survey, with a length of 303 feet and displacement of 3,805 tons.

52. How does an oceanographic ship anchor to take observations in the deep ocean? Although most oceanographic observations are made without anchoring, oceanographic ships sometimes anchor in deep water for several hours, days, or weeks to measure subsurface currents or to obtain repeated observations in one spot. The weight of the anchor need not be great, because the weight of cable lying on the bottom may be more than 2 tons. The U.S. Navy often uses anchors of 500 or 800 pounds, but Danforth anchors of only 40 pounds have been used to anchor in water a mile deep. In depths of 3,000 fathoms, wire tapering from $\frac{5}{8}$ to $\frac{1}{2}$ inch is normally used to lower the anchor. In greater depths, the taper is from $\frac{3}{4}$ to $\frac{3}{8}$ inch.

Free-fall anchors have been used for rapid anchoring in deep water. For example, the total elapsed time for planting a 4,000-

pound anchor at 17,250 feet was 16 minutes. No attempt was made to recover the cable and anchor. Even if a suitable winch had been available, the cost would have exceeded the value of the anchor and cable.

To prevent a ship from swinging on its mooring, it may be anchored fore and aft or it may tie to a bridle arrangement of three or four anchored buoys.

53. What are some of the more unusual oceanographic platforms from which observations are taken? Marine scientists are innovators and often resort to unorthodox ways to obtain observations. Ships are the usual platform but many other types have been used, particularly in recent years to obtain desired data at various locations. Lightships and weather ships have provided long strings of observations for a particular location. In the early 1950s the first oceanographic observations were being taken from offshore towers; this procedure is no longer unusual. For two decades or more, scientists have occupied floating ice islands in the Arctic to obtain marine and other geophysical observations for many successive months and years. Sometimes these ice islands are evacuated because of the deterioration of the ice structure or because the island has grounded, stagnated, or reached a position where additional observations are not considered worth while. Airplanes, helicopters, and satellites have all been used for obtaining oceanic data. Beneath the sea, submersibles of all types, both manned and unmanned, large and small, made of glass, steel, and aluminum have served as useful observation platforms. Undersea structures such as the *Sealab* and *Tektite* constructions are good platforms from which to obtain data. Each might be considered unusual or unique when first used but soon each becomes accepted as a standard platform. Perhaps the most unique are the flip-type ships which are floated to a location, then flooded to an upright position, resembling a giant spar buoy that extends more than 300 feet into the water. An unusual platform used in early 1970 for taking some observations in the Gulf of Mexico was a blimp. It was used to take airborne observations while an oceanographic ship took observations at the surface.

54. Do any oceanographic research ships work regularly in the Antarctic Region? A number of U.S. scientists conduct oceanographic investigations from two ships which operate each year in Antarctic waters. One, the *Eltanin,* a former cargo ship modified for oceanographic work, has been operated for the National Science Foundation since 1962 by the Navy's Military Sea Transportation Service. A second ship, the R/V *Hero,* a 125-foot trawler owned by the Science Foundation, was built in 1968 specifically to serve as a marine platform in Antarctic waters. Its home port is Palmer Station.

U.S. Coast Guard icebreakers are equipped with a suite of oceanographic equipment and instruments and usually carry several scientists and Coast Guard specialists who make oceanographic and ice investigations of the Southern Hemisphere's high latitude region. In the 1969–70 seasons the U.S. icebreakers working the area were the *Edisto, Glacier,* and *Burton Island.* Ships of other countries taking marine observations in the 1969–70 season included the Russian ship *Ob,* the icebreakers *General San Martin* from Argentina and *Fuji* from Japan, and New Zealand's HMNZS *Endeavor.*

55. What is Operation DEEP FREEZE? Operation DEEP FREEZE is a regular U.S. Navy operation conducted to supply instrumentation, equipment, personnel, food, fuel, and the other necessities of life by ship and by air to the remote scientific stations in Antarctica. It includes construction of buildings and carrying out the transfer of U.S. and, in many cases, foreign scientists and technical personnel. Operation DEEP FREEZE began in February 1955 when the U.S. Navy was assigned responsibility for providing support for U.S. participation in the Antarctic phase of the International Geophysical Year. A special group, the U.S. Naval Support Force, Antarctica, was established as part of the Atlantic Fleet to perform these logistic functions. The IGY proved so successful that a long range U.S. Antarctic Research Program was established and the Navy has continued to support the U.S. scientists working in the Antarctic. A rear admiral commands the Naval Support Force which includes skilled Army, Air Force, and Navy specialists as well as Coast Guard icebreakers; it is designated Task Force 43.

56. When was the first marine chart made? Most marine historians credit Marinus of Tyre with the construction of the first marine chart during the second century A.D.

57. When was the first chart prepared that showed the American Coast? There is some evidence that crude charts showing the North American coast may have been produced by Norse seamen before the time of Columbus. Nevertheless, the *Cosa Chart of 1500* is generally considered to be the earliest chart on which the American Coast is shown. It is drawn on oxhide in bright colors and purports to show the entire world. Asia and North America are shown as one continent, as the Pacific was unknown at that time. The chart was prepared by Juan de la Cosa, who accompanied Columbus as master of his flagship on the first voyage to America. He served as cartographer on the second voyage to the new world.

58. When were the first pilots produced covering the U.S. coast? Shipwrecks were frequent along the coasts of the United States in the early days of settlement due in part to the lack of charts and pilot information. During the colonial period, the British Government prepared a few vague charts of sections of the coast. Later, an atlas called *Atlantic Neptune* was issued; it was made up of reproductions of charts prepared and collected for British fleet use during the Revolutionary War. In 1730, Captain Southack of Massachusetts produced a *Coasting Pilot* which consisted of eight charts covering the New England Coast from New York northward. Even though it had many inadequacies, it was used for many years because there was nothing better. The *English Pilot* of 1774 and 1794 had a few improvements. In 1796 the *American Coast Pilot* was published by the Blunt Company and was popularly known as *Blunt's Coast Pilot*. It contained charts and much valuable written material collected from British work, private reports, and observations of seamen. This was undoubtedly the best coast pilot available for U.S. shores prior to 1800.

59. What is "Chart datum"? *Chart datum* (also called *sounding datum*) is the tidal datum used on nautical charts; it is the reference level from which soundings shown on the chart are determined. This notation is always carried on the chart, usually in the same corner as the scale.

60. What is meant by a Mercator projection? Gerhard Mercator (also known as Gerhard Krämer), a Flemish cartographer and mathematician, who lived in the sixteenth century, devised a technique whereby the surface of the globe could be projected and depicted on a plane surface as a map or chart. The Mercator projection is a conformal map projection in that it tends to preserve the correct shape rather than the correct size. One limitation is the expansion of latitude and longitude scales with higher latitudes. Its main advantages are simplicity in construction, convenience in plotting and scaling, and the fact that a straight line drawn in any direction is the shortest distance between two points. This is the reason that Mercator projections are preferred for nautical charts and navigation.

61. What does the scale on a chart denote? The scale of a nautical (or other) chart is the relationship that a measured distance on a survey, map, or chart bears to the corresponding actual distance on the earth. For example, if 1 inch on a chart corresponds to 1,000 feet (12,000 inches) along a shoreline, then the scale would be expressed as 1 inch = 1,000 feet. Often this is indicated on the chart as a ratio and is shown as 1:12,000.

62. What are isobaths? Curves of equal depth drawn on a chart to connect every point which is at the same depth level below the reference level (sounding datum) are called isobaths. On nautical charts, isobaths show the general configuration of the bottom and emphasize important navigational features such as shoals and channels. Areas between isobaths are often shown in color gradations of blue, the shoal areas being very light blue or white; for the deeper areas, a more intense blue is used. Depth contours are comparable to land elevation contours and the same principles apply.

63. What is the length of a *nautical mile*? The *nautical mile* (which is also called *sea mile* and *geographic mile*) is a unit of distance used in marine navigation. It is defined as equal to one sixtieth of a degree of a great circle on the earth. It is equal to a minute of arc measured along the equator or a minute of latitude on the chart being measured. On July 1, 1954, the United States adopted the use of the *international nautical mile,* which is 1852.0 meters (or 6,076.10333 feet). In practice the nautical mile is commonly defined as 6,080 feet; in Great Britain, this distance is sometimes referred to as an *Admiralty mile*. A nautical mile is 1.151 times as long as the statute or land mile.

64. What is a league? A league is a unit of distance which was commonly used 200 to 300 years ago. The *marine league* applies to distances over water and is equal to 3 nautical miles.

65. Where is the Intracoastal Waterway? This protected navigable channel runs almost the entire length of the east and Gulf coasts of the United States. Natural waterways, some of which have been deepened by dredging, are supplemented by connecting canals at needed intervals. The route extends through New Jersey; from Norfolk to Key West; across Florida from St. Lucie Inlet to Fort Myer and Tampa Bay; and from Carabelle, Florida, to Brownsville, Texas. The Intracoastal Waterway is used for commercial vessels as well as pleasure craft. Charts of the waterway are produced by the National Ocean Survey for the use of small-boat operators and yachtsmen.

66. What is meant by improved channels? These are dredged channels which are maintained at the assigned depths for the benefit of commercial shipping. The U.S. Army Corps of Engineers is responsible for both the initial dredging and maintenance of the assigned channel depth. On nautical charts, improved channels are shown by a black dashed line with channel depth and other navigational information shown on the chart or its margins.

67. What is dead reckoning? Ship position is determined by dead reckoning procedures whenever more precise positioning of the ship by other methods cannot be used. It is a procedure whereby the ship's location is determined by applying the ship's run to the last well-determined position, using the course (or direction) steered and the distance traveled as recorded in the log.

68. What does the pilot of a vessel do? He directs movement of a vessel through *pilot waters,* which are usually coastal areas approaching ports or channels leading to ports. Usually the pilot is one who has demonstrated extensive knowledge of channels, aids to navigation, and dangers to navigation and has an intimate knowledge of the particular water area for which he is licensed.

69. What is a chronometer? A chronometer is a portable timepiece with compensated balance which is capable of keeping time with extreme precision and accuracy. Mariners had long known how to determine positions of latitude; the chronometer permitted them to determine positions of longitude by keeping accurate time while at sea for the Greenwich meridian (0° longitude). The mariner by using his sextant and astronomical tables could determine the local time; the difference in time between the Greenwich meridian time and the local time (where 1 hour = 15 degrees) is used to determine the east-west position or longitude.

70. Who invented the chronometer? Because of a British naval disaster in 1707 in which many lives were lost through a navigational error, the Parliament offered a prize of from 10,000 to 20,000 pounds for development of a method to determine longitude accurately. John Harrison, the son of a British carpenter, built the first successful chronometer. His first instrument was tried at sea in 1735 and had an error of 3° of longitude. (This model is still in operation in the National Maritime Museum at Greenwich, England.) Improvements followed and his fourth model was tested in 1761 on a two-month trip to Jamaica. The time error was 9 seconds and the longitudinal error less than 2 minutes of longitude. Yet he received only a fourth of the prize money and the chronometer had to undergo yet another test. A copy of the fourth model was used by Captain Cook on his second expedition and proved an invaluable aid to his exploration.

71. What is Shoran? Shoran is a type of radar, a pulse-type electronic ranging system, developed during World War II for precision navigation and positioning of aircraft on bombing missions. It has been adapted for hydrographic and oceanographic surveying and provides a very accurate location of the survey ship's position. It is a line-of-sight system and for practical purposes is limited to distances of less than 75 miles and for use in coastal surveys. This term is an acronym derived from the words *sho*rt-*ra*nge *n*avigation.

72. What are the principal oceanographic organizations in the United States? Two decades ago the number of organizations actively engaged in oceanographic work might have numbered about 75 or perhaps 100 at most. During the 1960–70 decade, there was a great attraction to marine sciences work; many institutions and commercial organizations saw new opportunities in oceanic work and the number of government, educational, and industrial organizations engaged in oceanography increased many times. By 1970 the number probably reached several hundred. In 1970 there were some 22 or 23 federal agencies engaged in marine science activities; most authorities would list these as the principal activities: U.S. Naval Oceanographic Office (Suitland, Maryland), National Ocean Survey (Rockville, Maryland), National Marine Fisheries Service (Washington, D.C.), U.S. Coast Guard and Smithsonian Institution (Washington, D.C.).

Most of the thirty coastal states are conducting some marine work along their shoreline; Florida, Hawaii, North Carolina, and California have several very active state marine agencies. Among the educational and nonprofit institutions, Woods Hole Oceanographic Institution (Woods Hole, Mass.) and Scripps Institution of Oceanography (La Jolla, Calif.) still head the list. One might add another two dozen or more which have in recent years made noteworthy research contributions; Texas A&M (College Station, Texas), Oregon State University (Corvallis), University of Rhode Island (Kingston), University of Washington (Seattle), Johns Hopkins (Baltimore), and University of Miami (Florida) are good examples.

There are several hundred industrial organizations contributing to the marine sciences. Many of them offer very specialized services such as diving; others offer a broad range of engineering, research, testing, and analytical capabilities. Activities such as the Westinghouse Laboratories (Annapolis, Md.); Bissett Berman (San Diego), Sippican (Marion, Mass.), Perry Cubmarine Industries (Florida), Continental Shelf Data Systems (Denver, Colo.) are considered representative of the various commercial capabilities.

73. How was Woods Hole Oceanographic Institution (WHOI) established?

In 1884, Woods Hole, Massachusetts was selected as the site for the U.S. Commission of Fish and Fisheries' New England research laboratory. That location was selected because of its clean water, the absence of large rivers that would reduce salinities, and its deep-water access. Another facility, the Marine Biological Laboratory, was established and incorporated in 1888. It offered advanced course work and appropriate research facilities for individuals to conduct their own projects and investigations, usually in the summer.

The Woods Hole Oceanographic Institution was established as the result of a study and recommendations conducted by the National Academy of Sciences' Committee on Oceanography. Their study took over a year to complete; it began in October 1928 and their 163-page report was submitted on November 18, 1929. The Institution was incorporated on January 6, 1930, at which time the highly respected Harvard scientist, Dr. Henry B. Bigelow, was chosen as the first director. In February 1930 the new organization received from the Rockefeller Foundation the funds required to begin operation. A total of two million dollars was provided for buildings, boats, equipment, and other essentials and a half million dollars was allocated for operating expenses for the next ten years. Dr. Bigelow served as director until 1940 and was followed by Columbus O'D. Iselin, 1940–50 and 1956–58; Edward H. Smith, 1950–56; and Paul M. Fye, 1958 to present.

74. When was the Scripps Institution of Oceanography founded?   If a founding date were to be designated, 1903 would probably be a logical choice because this was the year in which activities of the University of California Summer Field Station were transferred to San Diego and became a function of the Marine Biological Association of San Diego, sponsored by E. W. Scripps and Ellen Browning Scripps. The field program investigations of life in the Pacific Ocean conducted by the University of California's Professor W. E. Ritter were started in 1892. The move to San Diego proved to be timely, as it established a permanent arrangement for conducting work. Funds for physical facilities and support for research work soon followed. In 1912 the Marine Biological Association was integrated into the University of California as the Scripps Institution for Biological Research. On October 13, 1925, the name was officially changed to Scripps Institution of Oceanography. Dr. W. E. Ritter, who conducted the first summer field investigations in 1892, served as director until 1923. He was followed by Dr. Thomas Wayland Vaughan, who headed the institution until 1936. The renowned Norwegian oceanographer Dr. H. U. Sverdrup was the director from 1936 to 1948, followed by Dr. Carl Eckart, 1948–50; Dr. Roger Revelle, 1950–64; Dr. Fred Spiess, 1961–65; and Dr. W. A. Nierenberg, who has served as director since 1965.

75. How many universities and colleges offer courses in oceanography? According to the data compiled in 1969 by the National Council on Marine Resources and Engineering Development, there are 83 U.S. colleges and universities offering a full marine science course of study leading to a degree. Another ten schools offer a few introductory oceanographic courses. Most universities have placed the marine science curriculum at the graduate level but there are some exceptions. The University of Washington offers an impressive curriculum for the student who wishes to obtain a bachelor's degree with a major in oceanography. Most other schools offering the A.B. or B.S. include one or more oceanographic courses as part of the requirements for a major in marine biology, meteorology, geography, or geology. In 1970 there were schools offering a full marine science curriculum in 25 of the 50 states plus the District of Columbia, Guam, and Puerto Rico. Some of the notable institutions which offer graduate degrees in oceanography are University of Alaska, Duke University, University of Hawaii, Johns Hopkins University, University of Miami, Oregon State, Scripps (U. of Calif.), Texas A&M, and Woods Hole Oceanographic Institution.

About ten community colleges scattered around the country offer two years of work with concentrated study in marine sciences; successful completion of such a course qualifies one as a marine technician.

76. When were the Sea-Grant Colleges established? The National Sea-Grant College and Program Act, Public Law 89-688, was signed by President Lyndon B. Johnson on October 15, 1966. It was modeled after a similar law passed more than a hundred years earlier authorizing establishment of Land-Grant Colleges to develop resources of the land through research, training, and extension. The new Sea-Grant Act was intended to assist institutions of higher education in taking a more active role in the development of marine resources.

77. Where is the Arctic Research Laboratory? On August 6, 1947, the first seven staff members arrived at Point Barrow, Alaska, to establish the Arctic Research Laboratory (ARL). Sponsored by the U.S. Navy and operated by the University of Alaska, it has developed into one of the world's foremost laboratories for conducting environmental research on high latitude phenomena. The ARL has complete laboratory and support services for researchers and scientists specializing in coastal oceanography, plant ecology, geomagnetism, coastal morphology, marine biology, polar meteorology, ionospherics, and ice and snow research. ARL scientists established ice island ARLIS as a drifting Arctic Ocean laboratory which served as a platform for making meteorological and oceanographic observations from April 1959 to May 1965.

78. Where in Antarctica is marine science work conducted?
The United States established and operated seven stations in Antarctica during the International Geophysical Year 1957–58. In 1970, four U.S. stations were still operated on a year-round basis by the Navy. These are the McMurdo and Palmer Stations on the coast and the Byrd and Amundsen-Scott South Pole Stations inland. Two other stations, Hallett and Brockton, are used in the summer only. McMurdo has grown in importance over the years and has become the main logistic center; it has also become the location from which much of the marine science work is done. Studies of the deep-diving characteristics of the Weddell seals, life cycle and population studies of seals, various types of fish studies and penguin studies are accomplished in the nearby area. Additional studies of penguins, algae, and other marine organisms were in progress during the summer of 1969–70 at the Hallett Station, only a short distance up the coast from McMurdo. Other marine biological, geological, and oceanographic investigations are carried on aboard research ships like the R/V *Hero* and at the U.S. Palmer Station, which serves as its home port.

III. THE CONTINENTAL MARGIN

79. What does the continental margin include? The *continental shelf* is a continuation of the geological and topographical features on the adjoining land. At the edge of the shelf, the *continental slope* leads downward to the deep ocean. At the bottom of the slope an area of deposition, known as the *continental rise* may form a gentler slope. Other features that may be present in the continental margins are trenches, ridges, and submarine canyons.

80. What is the continental shelf? In 1953 an international commission defined the continental shelf in this manner. "The zone around the continent extending from the low water line to the depth at which there is a marked increase of slope to greater depth. Where this increase occurs, the term (continental) shelf edge is appropriate. Conventionally, the edge is taken at 100 fathoms (or 200 meters), but instances are known where the increase of slope occurs at more than 200 or less than 65 fathoms. When the zone below the low-water line is highly irregular and includes depths well in excess of those typical of continental shelves, the term continental borderland is appropriate."

81. How was the continental shelf formed? Eighteen to twenty thousand years ago, continental glaciers stored up so much water that sea level was hundreds of feet lower than today. At that time, the continental shelf was part of the land area. Melting of the ice submerged the shelf. At one time the shelves were thought to be wave-cut terraces. Later they were considered depositional features. Evidence from coring operations does not fit either theory completely. Perhaps erosion was the main cause for some shelf areas and sedimentation for others. Or perhaps a combination of erosion and sedimentation is the explanation.

82. How far from the coast does the continental shelf extend? According to Dr. K. O. Emery, well-known Woods Hole geologist, the width of the continental shelf ranges from zero to

1,500 kilometers (about 940 miles). On a global basis, the continental shelf has been calculated to average about 78 kilometers (about 50 miles).

83. Does the depth of 100 fathoms determine the continental shelf?   While 100 fathoms (600 feet or 200 meters) is commonly used as a feature to determine the extent of the continental shelf, it is not in fact very accurate. On a close examination of the continental shelves of the world one will find the shelf edge at depths ranging from 10 fathoms to 275 fathoms. The average depth of the shelf edge computed for the world is about 67 fathoms.

84. How steep is the continental shelf?   The average slope of the continental shelves of the world is only 10 to 12 feet per mile.

85. What is the continental slope?   Starting at a depth of about 400 feet, the continental slope extends outward to a depth of 2,000 to 3,000 meters (6,500 to 9,800 feet). In many areas the continental slopes are cut by submarine canyons, some of which are equal to the Grand Canyon.

86. Where is the steepest continental slope?   The steepest known slope is off the east coast of Ceylon, where 30-degree slopes may be found. The average slope is about 5 degrees.

87. How was the continental rise formed?   This feature at the base of the continental slope is composed of sediments deposited by currents flowing down the slope. It is an underwater counterpart of the alluvial fans at the base of a mountain range.

88. What are the principal resources of the continental shelf?   Most of the large fisheries of the world are located on the shelf. Oil, gas, and some sulfur are obtained from offshore drilling. Mineral deposits known to exist on the shelf include diamonds, gold, tin, magnetite, iron, chromite, titanium, thorium, and rare earths.

89. What was the significance of the Presidential Proclamation of September 28, 1945?   This proclamation, issued by President Harry S. Truman, set forth the policy of the United States with regard to the protection of fishery resources of the high seas adjacent to its coast and regarding it proper to establish conservation zones for those areas. Further, it served to extend jurisdiction and control of the United States over the natural resources of the subsoil and seabed of this continental shelf, but not including the waters above, which are considered as high seas.

90. Do all coastal states of the United States have the right to develop submerged lands an equal distance from shore? On May 22, 1953, H.R. 4198 of the 83rd Congress, 1st Session, was signed and became Public Law 31 and is often referred to as The Submerged Lands Act. It limits state jurisdiction over water-covered lands to 3 miles seaward of mean high tide. In a decision made on May 31, 1960, the Supreme Court upheld claims of Texas and Florida to a seaward boundary of 9 nautical miles for submerged lands in the Gulf of Mexico.

91. What is the Outer Continental Shelf Lands Act? This act, often referred to as Public Law 2-12, was passed by the 1st Session of the 83rd Congress and signed into law on August 7, 1953. It supplements Public Law 31 and provides for the jurisdiction, control, and administration by the United States over the submerged lands seaward of the states' boundaries as defined in Public Law 31; that is, over the outer continental shelf. It also authorizes the Secretary of the Interior to lease such lands for certain purposes.

92. What are submarine canyons? They are steep-sided valleys cut into the continental shelf and continental slope. Some extend outward from mouths of rivers; others have no relation to present-day rivers. A submarine canyon 4 miles wide and 240 feet deep extends from the mouth of the Ganges River into the Bay of Bengal for a distance of more than 1,000 miles. The Congo Canyon has been traced for 145 miles to a depth of 7,500 feet. Off the Hudson River a canyon extends 200 miles across the continental shelf. The continental shelf along the Pacific Coast of the United States is narrow and its canyons are deep and steep-sided. The Monterey Canyon, off California, is sometimes compared with the Grand Canyon.

93. How were submarine canyons formed? The origin is still under dispute. There is speculation that channels may have been cut by rivers while the land was above sea level, but many scientists are skeptical of this theory. Such an origin would imply that the coastal areas of the world have been depressed thousands of feet. The most popular theory is that the canyons were scoured out by turbidity currents.

94. What are turbidity currents?   These occur when sediments on the continental slope are dislodged by earthquakes and begin sliding down the slope. A current is created by the increased density of the sediment-laden water. This current, in turn, dislodges more sediment which continues downward at greater speed. The sediment-laden currents cause scouring of the sea floor; it is believed that they contribute to the flushing and erosion of submarine canyons. If the turbulence is sufficient to keep sediments in suspension, turbidity currents may flow for great distances; the sediments are finally deposited on the abyssal plain.

95. How has the speed of turbidity currents been measured?   Turbidity currents have broken off series of submarine cables; the time between cable breaks makes it possible to compute their approximate speed. If the slope is steep and long, the speed may reach 50 miles per hour.

96. How do submarine canyons aid navigation?   About 100 miles east of Cape Cod, there is a series of submarine canyons cutting into the continental slope off Georges Bank. These canyons cross the major shipping lanes between New York and Europe. The canyons have been accurately surveyed by ships of the National Ocean Survey of the National Oceanic and Atmospheric Administration, and are shown on navigational charts. As ships approach the area, they normally switch on the echo sounder. When the ship crosses the first canyon, the navigator checks the maximum echo sounder depth against the chart. He can then determine at what point he crossed the canyon. As the ship crosses the other canyons, additional position checks can be made and speed of advance can be determined.

97. What information do ocean engineers need for offshore construction?   Construction on the continental shelf necessarily involves detailed knowledge of the areas under consideration—

prevailing weather, currents and tides, bearing capacities, earthquake and fault zones, sediments, the presence or absence of shipping lanes and commercial fishing grounds and underground cables and moorings.

98. What is erosion? According to riparian law, it is defined as the gradual and imperceptible washing away of the land along the sea by natural causes. The term also is applied to submergence of land due to encroachment of the waters. Though this definition is in general accepted by most, the term *erosion* is frequently applied to beach and shore destruction that has been accomplished over a very short interval of time by a severe storm.

IV. MARINE GEOLOGY

99. Who made the first recorded observation of the sea bottom?   In the writings of Herodotus, which were made about 450 B.C., we find one of the earliest, if not the first, records of information about the bottom sediments of the Mediterranean off Egypt. He wrote, "The nature of the land of Egypt is such that when a ship is approaching it and is one day's sail from the shore, if a man try the sounding, he will bring up mud even at a depth of 11 fathoms."

100. What is the source of sediments?   The most important source is earth and rock material carried to the sea by rivers and streams; the same materials may also be transported by glaciers and winds. Other sources are volcanic ash and lava, shells and skeletons of organisms, chemical precipitates formed in seawater and particles from outer space.

101. What is the most common sediment?   Mud is the most common sediment on both the deep-ocean floor and the continental shelf. The term *mud* is properly applied to silt and clay-sized particles less than 0.06 millimeter, although it is often applied to any sticky fine-grained sediment.

102. What sediment is found in deep-ocean basins?   Red clay is found at depths greater than 4,500 meters (14,764 feet). Red clay is the sediment that remains after the carbonate has been dissolved away. At this depth, seawater is more corrosive and calcareous sediments cannot survive.

103. What are calcareous sediments?   They are the remains of organisms which concentrate calcium carbonate. Globigerina ooze is the major calcareous sediment in moderately deep water, to depths of 4,500 meters.

104. What is siliceous ooze?   This is a fine-grained sediment containing the remains of single-celled radiolarians (animals) and diatoms (plants), both of which are rich in silica. The organisms thrive in areas where upwelling of cold water brings nutrients to the surface. Belts of siliceous ooze are found along the Antarctic continent and in areas of the northern and eastern Pacific.

105. What are terrigenous sediments? These are deposits of debris eroded from land areas and carried to sea. The deposits, mainly silts and clays, are usually found on the continental shelf.

106. What are glacial deposits? Glaciers and icebergs transported sediments and rock fragments during the ice ages and deposited them on the ocean bottom as the ice melted and retreated. They are most common on the continental shelf in the northern latitudes.

107. What are volcanic sediments? Pumice and ash, resulting from volcanic eruptions, are found as sediments throughout the oceans of the world in both deep and shallow water.

108. What are pelagic sediments? These are sediments that have been carried in suspension in seawater, distributed widely and deposited slowly. The fine particles of red clay found in deep ocean basins are pelagic sediments, as are the calcareous and siliceous remains of planktonic organisms found in deep water. Volcanic ash, dust, and chemical precipitates are also present in pelagic sediments.

109. How are sediments transported? Streams and currents are the main agents of transport for land-derived sediments. Wind can carry fine particles for great distances; particles from the Sahara are scattered widely over the Atlantic. Dust particles can remain in the atmosphere indefinitely unless removed by rain. Icebergs drifting to lower latitudes deposit coarse material and even boulders as they melt.

Once sediments are deposited they may be stirred by waves and turbulent motion and moved by currents.

110. How thick are the sediments on the ocean floor? Thickness of the sediments which cover the ocean floor may vary from almost nothing to as much as 4,000 meters (13,123 feet); in trenches the thickness can be much greater. The average of the world's oceans is about 300 meters (984 feet) thick. Sediments in the

Atlantic vary between 500 and 1,000 meters (1,640 and 3,281 feet), averaging about 750 meters (2,460 feet). Because much of the Pacific is far from the land which is the source of sediments, thicknesses are less, averaging between 300 and 600 meters (984 and 1,968 feet). Basins of the Indian Ocean have about the same thickness as the Pacific.

111. Why aren't the sediments thicker?   If the rate of sedimentation had been the same through geologic time as it is today the sediments would have been many times thicker. Either the rate was slower in the past or the time of accumulation is far less than the age of the oceans.

112. How is sediment thickness measured?   Thickness of sediment layers is measured by seismic reflection profiling. A small bomb is exploded just under the surface of the sea to produce a sound source. Part of the sound is reflected from the ocean bottom and part is reflected from the sediment and rock strata. The time difference in return of echoes indicates thickness. More than one million miles of profile data have been obtained by this method, furnishing information on the major sediment features of the world's oceans.

113. How fast do sediments accumulate?   Red clay in very deep water accumulates very slowly, the rate being as low as 1 millimeter in 1,000 years. Probably the most accurate method of dating sediments is the radiocarbon method. Rates calculated by this method are 4.3 mm per 1,000 years in the Pacific and about twice that rate in the Atlantic. Carbonate sediments accumulate faster than any other major type, about 1 to 4 centimeters per 1,000 years.

114. How are bottom sediments sampled?   Depending on the character of bottom, depth and nature of the investigation, oceanographers may use corers, snappers, grabs, or dredges. The sampler must penetrate to the required depth and retain the sample while being raised from the sea bottom. Corers are used to obtain undisturbed samples of sediments. If the bottom is rock, coral, or gravel, a coring tube may be damaged or fail to penetrate; for such areas grab samplers and dredges are used.

115. What is a grab sampler? The orange-peel sampler is commonly used for obtaining samples of sand, gravel, coral, and shell. It is named for its resemblance to the skin of an orange peeled in four sections. The weight of the sampler causes it to penetrate the bottom. As it is retrieved the jaw closes, retaining the sample. One disadvantage is that mud may be washed out of the sampler as it is raised through the water; a canvas cover reduces the loss somewhat.

116. How are snappers used? Samplers of the clamshell type are spring-loaded and obtain surface samples when their jaws are triggered by impact with the bottom. Like the orange-peel sampler, there is a tendency for mud to wash out while the sampler is being hauled to the surface.

117. How are dredges used? When corers and grab samplers fail to obtain samples, dredges are usually employed. Heavily constructed box-shaped dredges with chain mesh bags are used to break off and collect pieces of solid rock. Smaller cylindrical dredges are sometimes used to collect nonconsolidated material in shallow water. Both types are towed by a ship.

118. What are bottom corers? These are pipes that are driven into the ocean floor by their own momentum. Several hundred pounds of weights may be added to increase penetration. Plastic tubes are often used as liners so that the core can be preserved and returned to a laboratory for study.

One of the simplest sampling devices is the *Phleger corer,* having a length of three feet and barrel diameter of one and one half inches.

119. How are undisturbed cores obtained? To overcome friction and distortion caused by the coring tube being forced into the sediment, a piston can be added inside the tube; one such device is the *Kuhlenberg corer.* The piston is attached to the lowering wire in such a way as to remain almost stationary just above the sediment surface while the tube is penetrating. This creates a partial vacuum and reduces compression and distortion of the sediment core.

120. How long are the longest cores? By adding as much as 1,500 pounds of weight to a long-barreled corer, cores as long as 25 meters (82 feet) have been taken.

121. What is a boomerang corer?   To avoid the time-consuming process of lowering a corer several miles on a cable, scientists at Woods Hole Oceanographic Institution have developed a free-fall corer with no line attached. Cast-iron ballast carries the device rapidly to the bottom. The weights remain on the bottom and the core is buoyed to the surface by glass floats.

122. How are deep holes drilled in the ocean bottom?   Oil drilling techniques long used on land have been taken to sea for drilling undersea oil wells and for obtaining sediment cores for scientific study. Thirty-foot sections of 3-inch pipe are coupled together to form a string from the ship to the sea bottom. A rotary diamond bit cuts into the bottom. A core barrel is lowered through the drill pipe to take 20-foot core samples.

123. What has been learned from deep-sea drilling?   One important finding was that the ocean basins appear to be much younger than the earth; no sediments older than 140 million years have been found. Strong evidence has been found to support sea-floor spreading and continental drift. Proof has been obtained that oil and gas occur in deep-sea sediments. It has been established that the Northwest Pacific existed long before the Atlantic Ocean. Many more discoveries are being made as the cores are studied.

124. What new techniques are being used to study sediments?   Use of X-ray diffraction and fluorescence and infrared and electron-microscopes has greatly increased knowledge of the mineralogical composition of marine sediments.

125. Why is the mineralogical character of sediments important?   The factors that give a sediment its mineralogical character are intimately related to processes such as wind patterns, ocean-water movements, climatic conditions, and volcanic activity. Evidence of processes that took place in the geologic past are recorded in the marine sedimentary column.

126. What colors are found in sediments?   White and light shades are usually associated with coarse-grained quartz or limestone deposits. Darker shades of red, blue, and green are usually found in muds containing iron or manganese. Black mud is often found in an inlet or depression cut off from current flow.

127. What are the major features of the ocean bottom?   All oceans have *continental margins, ocean basin floors,* and *major ridge systems.* The continental margin consists of the *continental shelf* and the *continental slope.* In addition to the major ridge system, additional ridges and rises occur throughout the ocean basins.

128. How much of the ocean floor has been charted?   Not more than 5 percent of the world's ocean floor has been charted with any degree of reliability, and most of this was done during the International Geophysical Year (1957–58).

129. When was the first deep-sea sounding made?   The first recorded deep-sea sounding was made on January 3, 1840, by HMS *Erebus* enroute to the Antarctic. Hemp line was lowered from one of the ship's boats and the rate of descent was measured to determine when bottom was reached. *Erebus* reported a depth of 2,425 fathoms at 27° 26′ S, 17° 29′ W. In 1968 the Coast and Geodetic Survey Ship *Discoverer* sounded the same spot with an echo sounder and recorded 2,100 fathoms. A depth of 2,400 fathoms was discovered just 3½ miles away.

130. What is a fathom?   A fathom is 6 feet or 1.83 meters. The word comes from the Old English and means *outstretched arms.* In the days of sailing vessels, soundings were made by lowering a lead weight on hemp line. As the line was retrieved it was measured with outstretched arms.

131. How are ocean depths measured?   Until about 1920 all soundings were made by lowering a weight attached to hemp line or wire. Even using this tedious method, enough soundings were made to indicate the general features and depths of the ocean bottom.

In 1927 the German vessel *Meteor* used an echo sounder to determine the bottom topography of the South Atlantic. The detailed information obtained by *Meteor* proved beyond doubt that the sea bottom is not a featureless plain.

Continuously recording echo sounders have been available since

World War II. During a recent twelve-month period the Naval Oceanographic Office collected and processed more than 220,000 nautical miles of bathymetric survey data.

132. What is the average depth of the oceans?   Recent estimates have ranged from 11,660 to 12,230 feet. Only about 10 percent of the ocean is deeper than 18,000 feet. About 7 percent of the ocean is less than 600 feet deep.

133. What are ocean basins?   Boundaries fall within the depth range of 3,000 to 6,000 meters (9,842 to 19,685 feet). Only about 1 percent of the ocean floor has a depth greater than 6,000 meters.

134. What are abyssal plains?   These are the extremely smooth plains found in the ocean basins, generally at depths below 3,700 meters (12,139 feet). The plains slope only a few feet in a hundred miles.

135. What are trenches?   They are narrow, often arc-shaped depressions in the ocean floor. Width may be 50 to 60 miles and length as much as a thousand miles. They are the deepest parts of the oceans, with depths from 20,000 to more than 36,000 feet.

According to the late Dr. Anton Bruun, Danish oceanographer, trenches cover an area equal to nearly half of Europe. In other words, trenches occupy about 1.8 percent of the total area of the ocean floor, or about 1.3 percent of the earth's surface.

136. What are the major trenches of the world oceans?   Trenches are found in each of the three oceans; the Pacific Ocean has seven major trenches, the Indian Ocean has four and the Atlantic Ocean has two. These are:

In the Pacific— Marianas Trench, Tonga-Kermadec Trench, Japan-Kuril Trench, Aleutian Trench, Philippine Trench, Middle America Trench, and the Peru-Chile Trench

In the Indian— Java Trench, Diamantina Trench, Mauritius Trench, and the Vema Trench

In the Atlantic—Puerto Rico Trench and the South Sandwich Trench

137. What is the origin of trenches? Trenches are associated with areas of frequent volcanic and earthquake activity. These are areas where the earth's crust is thin. Some geologists believe that island arcs and trenches are formed by spreading of the sea floor outward from the mid-ocean ridges and the resultant wrinkling by compression against the continental borders.

138. Where are some of the very deep spots in the ocean? The ten deepest soundings in the ocean have all been located since the end of World War II. Widespread use of electronic echo-sounding equipment aboard all ships during that conflict contributed greatly to these discoveries. The ten deepest soundings are all recorded in the Pacific Ocean:

Depths (feet)	Ship and Country	Date	Place
36,198	*Vityaz* (USSR)	1957	Marianas Trench— Challenger Deep near Guam
35,800	*Trieste* (U.S.)	1960	Marianas Trench— Challenger Deep
35,702	*Vityaz* (USSR)	1951	Tonga Trench—north of New Zealand
35,631	*Challenger* (U.K.)	1951	Marianas Trench— Challenger Deep
35,598	National Geographic Society (U.S.)	1965	Tonga Trench
35,460	*Challenger* (U.K.)	1951	Marianas Trench— Challenger Deep
34,884	*Horizon* (U.S.)	1952	Tonga Trench
34,587	*Vityaz* (USSR)	1954	Kuril Trench, near Japan
34,578	*Galathea* (Denmark)	1951	Philippine Trench
34,062	*Vityaz* (USSR)	1953	Kuril Trench

139. Where have the deepest soundings been made in the Atlantic Ocean? Soundings taken in the Puerto Rico Trench are the deepest for the Atlantic Ocean. It is almost 5 miles deep wherever a sounding is taken and extends some 450 miles. The trench bottom has an abyssal plain 150 miles long. Some depth measurements taken for the Puerto Rico Trench are:

Depths (feet)	Ship and Country	Date	Place
30,210	USS *Milwaukee* (U.S.)	1941	Puerto Rico Trench
28,374	USS *Archerfish* (U.S.)	1961	"
27,978	USS *Rehoboth* (U.S.)	1955	"
27,876	"	1955	"
27,810	USS *San Pablo* (U.S.)	1955	"

140. Why are there no trenches along the U.S. coast? There is evidence from seismic refraction measurements that trenches border the continental slope off New England, but they have become filled with sediments. Trenches may at one time have existed at the boundaries between all continents and oceans.

141. How are trenches filled with sediments? If rivers discharge sediments onto the continental shelf and slope, turbidity currents may carry them seaward until they are deposited in the trenches. Discharge from the Orinoco River has filled part of the Puerto Rico Trench to the extent that the trench can be traced only by seismic refraction and gravity measurements.

142. What are mid-ocean ridges? There is a belt of ridges passing through the Atlantic, Pacific, Indian, and Arctic Oceans for a distance of more than 40,000 miles. In many places the ridges extend more than 16,000 feet above the surrounding ocean basins.

143. When were mid-ocean ridges discovered? In 1873 scientists of HMS *Challenger* discovered the Mid-Atlantic Ridge while sounding with a lead weight on a hemp line. Until that time there was no knowledge of the great system of mid-ocean ridges. During the period 1925 to 1927 the German ship *Meteor* made a detailed study of the ridge with an echo sounder.

144. What is the Mid-Atlantic Ridge?   This ridge extends in a curving line for about 15,000 miles midway between the American and European-African continents. Its width is 600 to 1,000 miles and it rises as much as 9,000 feet above the surrounding ocean basins.

145. How was the Mid-Atlantic Ridge formed?   There are many theories. It may be a block thrust upward between major fractures, or a chain of volcanic extrusions or an area of folded rocks. The true geologic nature is still unknown.

146. What is the rock material of oceanic islands?   Oceanic islands are built up by successive flows of lava. The Hawaiian Islands are part of a volcanic chain that extends across the Pacific for nearly 2,000 miles. Although nothing except coral may show at the surface, drilling has confirmed the underlying base of coral islands and atolls.

147. What is the definition of a reef?   A reef is a rocky or coral elevation dangerous to surface navigation which may or may not uncover at the sounding datum. A rocky reef is always detached from shore; a coral reef may or may not be connected with the shore.

148. Are all oceanic islands volcanic?   A notable exception is the Rocks of St. Peter and St. Paul in the middle of the Atlantic, just north of the equator. Charles Darwin noted their unique character during his voyage in the *Beagle* in 1831. These islands are composed of peridotite, an ultrabasic rock.

149. How many volcanoes exist on the sea floor?   Although only the largest volcanoes extend above the surface of the ocean, there are an estimated 10,000 volcanoes on the sea floor.

150. What is Surtsey?   On November 15, 1963, a new island appeared near the Westman Islands, just south of Iceland. The scientific world was alerted by the fishing vessel *Isleifur II,* which noted submarine volcanic activity in the area the previous day. Geologists and geophysicists from Iceland, Europe, and the United States have had a unique opportunity to photograph and observe the development of a permanent island from an undersea volcano. Initially, there was some doubt about the permanence of Surtsey because wave action eroded much of the new volcanic pumice and cinders, but when geologists observed the lava flow in April 1964 they were convinced that the island would not be washed away by waves. By 1965, Surtsey covered approximately 1¼ square miles and was growing at the rate of 1 acre each day. The reported lava discharge at that time was 180,000 tons per hour. This new island has also afforded biologists and ecologists an opportunity to observe the beginning of life on newly emerged land.

151. Are there any undersea volcanoes in U.S. territorial waters?   Most of the active volcanoes in the United States are in Alaska and the outlying islands of the Aleutian chain. According to the U.S. Geological Survey, there are 36 active volcanoes on the Aleutian Island Arc which includes the Alaskan Peninsula, but few have been the target of extensive study. Bogoslof Island is one active volcano that is known to have emerged and submerged in the sea more than once in historic time. Though undersea volcanic activity is not presently known along the Aleutian Islands, there is a distinct possibility that such events can occur in this area at any time.

152. What are seamounts?   Seamounts are relatively isolated conical peaks or groups of peaks which have been found in all oceans. To qualify as a seamount, a rise must be at least 1,000 meters (3,281 feet) above the surrounding plain.

153. How many seamounts are there?   At least 1,400 seamounts have been discovered in the Pacific Ocean. There are thought to be at least 10,000 seamounts in all the oceans of the world. In a 24-day survey in 1969 the National Ocean Survey Ship *Oceanographer* located 25 previously unreported seamounts in the South Pacific. Some rose as much as 10,800 feet above the ocean floor.

154. What is the origin of seamounts?   They are volcanoes, occurring as isolated peaks or as groups of peaks. Their distribution in linear chains gives credence to the theory that they are caused by fissure eruptions.

155. What is the Cobb Seamount?   This seamount is located in water 8,500 feet deep, about 270 miles off the coast of the state of

Washington. Its unique feature is that its top is a 20-acre plateau just 112 feet below the surface. The Cobb Seamount has been used as a site for anchoring research ships and instrumented buoys to collect oceanographic data.

156. What are guyots? In the Pacific Ocean many flat-topped seamounts, known as *guyots,* have been found. A Princeton University geologist, Dr. Harry H. Hess, discovered guyots in the examination of echo sounder records while in command of a naval vessel during World War II. Guyots are much less common in the Atlantic.

157. What makes guyots flat? Professor Hess, who discovered guyots, explained the flat tops as erosion at sea level. Because the tops of some guyots are 1½ miles below the surface of the sea, other explanations have been sought. It was thought that the seamounts were capped by sediments which made them flat. Dredging and sounding proved that they were mostly bare of sediments. It now appears that Dr. Hess was correct. The seamounts were eroded at sea level and they subsided because the sea floor was unable to support their tremendous weight. General downwarping of the sea floor may have accompanied the local subsidence.

158. What part do waves and currents play in modifying the sea floor? Although the effect of surface waves does not extend far below the surface, there are internal waves between layers of water thousands of feet below the surface. Photographs of the tops of seamounts show ripple marks of the type caused by waves.

Deep-sea cores contain clean sand and gravel, indicating the presence of currents at great depths.

159. What animal remains are found on the deep ocean floor? Shark teeth are very resistant to decomposition and are commonly found in deep-sea sediments. Ear bones of whales and fish bones are sometimes found. Very little skeletal material is found in sediments because the solubility of phosphate compounds increases with depth.

160. What is continental drift? Anyone looking at a world globe will notice that the eastern coastline of South America fits the western coastline of Africa rather closely. This was noted by Sir Francis Bacon in 1620. In 1912 the German scientist Alfred Wegener hypothesized that all continents once formed a single land mass which he called *Pangaea*. At the time of his death in 1930 his hypothesis had gained little acceptance.

161. Is there evidence that all continents were once joined? Geologists have correlated the rock structures of South America and Africa. In recent years oceanographers of the National Oceanic and Atmospheric Administration have found that there is a very close fit between the 1,000-fathom lines off the coasts of North America and North Africa if Cape Hatteras is linked to Cape Verde.

162. What is sea-floor spreading? This concept states that convection currents cause rock from the earth's interior to rise slowly over millions of years under the mid-ocean ridges and spread slowly along the sea floor, pulling the newly formed crust of rock along. At the continental margins the rock circulates downward.

163. How does the concept of sea-floor spreading support continental drift? The concept is that continents do not drift through the mantle but merely float on it as it moves away from rises and ridges. This concept has been much easier for scientists to accept than the original Wegener hypothesis of continental drift.

164. What is the evidence for sea-floor spreading? Deep-sea cores obtained by drilling between western Africa and South America revealed that the youngest rock materials were closest to the Mid-Atlantic Ridge. Age increased at the rate of a year for each 2 centimeters outward from the ridge. These measurements indicate that South America and Africa split apart at least 150 million years ago.

Additional evidence is being obtained by measurement of magnetic orientation in the rocks under the sea.

165. How was the Red Sea formed? If convection originates under a mid-ocean ridge, sea-floor spreading results. However, when convection originates under a continent, the outward spreading can split the continental crust. This appears to be what is happening in the Red Sea and the Gulf of California.

166. What causes variations in gravity? Measurements of gravity are an indirect way of weighing. Differences in the thickness and density of the underlying rock cause differences in gravity. Precise measurements of gravity at sea level and beneath the sea furnish valuable clues to the present condition and past history of the earth's crust. Data collected from all over the world, on land and sea, indicate that the ocean bottom consists of denser rocks than the continents.

167. When were gravity measurements first made at sea? In 1929, F. A. Vening Meinez, a Netherlands geophysicist, developed a pendulum gravimeter that is still used today. He made his measurements in a submerged submarine in order to have a stable platform.

168. When did U.S. scientists first measure gravity at sea? In 1931, Maurice Ewing and Harry H. Hess made precise gravity measurements aboard the submarine USS *Barracuda*.

169. When was gravity first measured from a surface ship? In November 1957, J. Lamar Worzel of Lamont (now Lamont-Doherty) Geological Observatory made the first successful gravity measurements from a surface ship.

170. How extensive are marine gravity surveys? During one twelve-month period, ending July 1968, 5 ships of the U.S. Naval Oceanographic Office collected gravity measurements along more than 180,000 miles of survey track. These data are used for refinement of inertial guidance and navigational systems.

171. What is the difference between earthquakes and seaquakes? Most earthquakes occur under the sea floor and the mechanics are the same whether occurring under a land mass or the ocean floor. Earthquakes ashore are sometimes called land quakes and those at sea are referred to as seaquakes.

172. Can a seaquake be felt on shipboard? Frequently the first waves to reach the sea surface are not strong enough to be felt on board a ship, and pass into the atmosphere to create a sound wave. When the frequency is in the audible range there will be a loud noise resembling an explosion. Normally there is no sign of disturbance on the sea surface, but severe vibrations of the ship's hull are caused by the wave front.

173. Can seaquakes damage ships?   Vibrations from seaquakes rarely damage ships because they are built to withstand severe external forces. However, on April 15, 1947, a ship off Point San Telmo, Mexico, reported that earthquake vibrations caused a well-secured deck cargo of heavy steel prefabricated construction sections to shift 6 inches.

174. Are underwater earthquakes the cause of most submarine cable breaks?   When Matthew Fontaine Maury wrote about the ocean bottom environment in 1855, knowledge available at that time caused him to conclude that cables laid on the ocean floor "would lie in cold obstruction, without anything to fret, chafe or wear, save along the tooth of time." Of course this was based on his concept that "agents which disturb the equilibrium of the sea . . . all reside near or above its surface; none of them have their home in the depths." By the beginning of the twentieth century, evidence was beginning to accumulate to indicate that Maury's information was incomplete and his conclusion about undersea cable endurance was wrong.

The Bell Telephone Laboratories have conducted and sponsored many studies on submarine cable breaks. Based on work done at the Lamont-Doherty Geological Observatory, over the past 20 years, Dr. Bruce Heezen concludes that by far the greater percentage of cable breaks occur in waters less than 200 fathoms (1,200 feet) deep with most of these failures being caused by "chafing and corrosion." When submarine cables are laid on steep slopes, and near unstable sediment sources such as river mouths, the cable breakage hazard is increased by an order of magnitude. Where cables cross significant fishing areas such as the Grand Banks, trawling activity is the major cause of cable breaks. Fouling by bottom organisms does not seem to impair cable operations to any substantial degree except in water depths less than 10 fathoms (60 feet). Undersea earthquakes, volcanoes on the deep ocean floor, and deep feeding animals cause some cable breaks but these do not represent a very substantial percentage of the total cable failures reported and investigated.

175. How is sediment thickness determined? A sound is generated just below the sea surface by a small explosive charge, compressed air, electric spark, or ignition of gas. The sound is reflected by the ocean bottom and by each subsurface layer of sediment. These measurements can be made from a ship moving at a speed of 4 to 6 knots; newer sub-bottom profilers will permit higher ship speeds. Measurements can be made to depths of 10,000 feet or more beneath the ocean floor.

176. What is the MOHO? This is the name commonly used for the *Mohorovicic discontinuity,* the boundary between the earth's crust and mantle. It was named for the Yugoslav seismologist who discovered its existence. The crust is the surface layer of rock, averaging 125,000 feet in thickness under the continents; beneath oceans it is only 15,000 to 20,000 feet thick. This is why the planned Mohole was to be drilled through the ocean floor. At the Mohorovicic discontinuity, the speed of earthquake waves changes abruptly, indicating a difference between rocks of the crust and of the mantle.

The objective of the now abandoned Mohole Project was to drill through the MOHO and obtain samples of the mantle rock. Some of the questions which led to the project are: How did the rocks of the oceanic crust become separated from the rocks of the continental crust? How was the crust differentiated? And from where did the layers of oceanic crust come?

Achievements that arose from the Mohole Project included ways to core the ocean bottom in deep water, a better understanding of the geophysics of several ocean areas, and improvement of drilling equipment and techniques.

177. When did magnetic surveying at sea begin? Edmund Halley, a British astronomer, made observations of magnetic variation in the Atlantic in 1698–1700. He prepared a chart which he hoped would enable navigators to determine their longitude by comparison with shipboard observations.

178. When did the United States begin collecting magnetic data at sea? The U.S. Exploring Expedition of 1838–42, under the command of Navy Lieutenant Charles Wilkes, began the collection of magnetic data on a small scale. In 1881 the worldwide collection of data was undertaken on a large scale when all wooden ships of the U.S. Navy were requested to report observations of compass declination throughout the oceans of the world.

179. How is magnetic intensity measured? A magnetometer is towed behind a ship at a distance so that the instrument is unaffected by the steel hull. The instrument commonly used on oceanographic surveys is the nuclear precession magnetometer. The principle of operation is that protons precess (spin) in a magnetic field. Frequency of their spin is proportional to total intensity of the magnetic field.

180. How are airborne magnetic surveys conducted? Since 1953 the Naval Oceanographic Office has been conducting airborne geomagnetic surveys of ocean areas. Aircraft generally fly along east-west lines approximately 200 miles apart. Well over 200,000 nautical miles of track are covered annually. Airborne surveying has permitted collection of data in polar regions which were previously inaccessible. Data are used to construct reliable navigational and world isomagnetic charts.

V. PHYSICAL PROPERTIES
OF SEAWATER

181. What physical characteristic is singularly unique for water? Water is a most unusual substance because it exists on the surface of the earth in its three physical states, solid, liquid, and gas (ice, water, and water vapor, respectively). There are other substances that might exist in a solid and liquid or a liquid and gaseous state at temperatures normally found at the earth's surface, but there are few if any substances which occur in all three physical states.

182. What are some important physical and chemical characteristics of water? First, water is odorless, colorless, and tasteless.

Second, it is the only substance known to exist in a natural state as a solid, a liquid, and a gas on the surface of the earth.

Third, it is the universal solvent. More salts and other substances are dissolved by water than any other material.

Fourth, water does not corrode, rust, burn, or separate into its component elements easily. It is virtually indestructible chemically.

Fifth, it can corrode almost any metal and wear away the most solid rock.

Sixth, a unique property of water is that it expands and floats on water when frozen (solidified).

Seventh, water has a great affinity for itself; the highest of all liquids. This is why water exists in the form of drops, a spheroid shape which has the least surface area. Surface tension is the characteristic responsible for the capillary action necessary to life of both plants and animals.

Eighth, water does not freeze at the temperature of greatest density 4° C but at 0° C.

Ninth, water has a capacity for absorbing great quantities of heat with relatively little increase in temperature. In addition, the latent heats of fusion (80 cal./gm) and vaporization (540 cal./gm) absorb substantial additional heat with no increase in temperature at the freezing point and vaporization point.

Tenth, water when distilled, is a very poor conductor of electricity but when salts are added even in small amounts it becomes a very good conductor of electricity.

183. What are the physical properties of seawater?   Physical properties which are commonly measured or computed are temperature, density, pressure, ice, water color, transparency, and sound velocity. Other properties, such as electrical conductivity, may be measured or computed for special studies. Salinity (a chemical property) is often included with the physical properties because it is so frequently measured along with temperature.

184. Why are temperature and salinity measured together? Temperature, salinity, and pressure (based on depth) are required for computation of density (weight per unit volume) and dynamic currents. Every area of the ocean has its own range of temperature and salinity; these properties, together with dissolved oxygen, help physical oceanographers trace water movement at various levels.

185. How are subsurface temperatures measured?   Temperatures are measured at discrete depths by reversing thermometers. Continuous records of temperature with depth are obtained by mechanical bathythermographs or by electronic instruments.

186. What is a deep-sea reversing thermometer?   This glass thermometer has been the basic oceanographic instrument for the past hundred years. It is built with a narrow constriction in the glass column just above the bulb. Reversing the thermometer breaks the column of mercury so that temperature at depth of reversal can be read when the thermometer is brought back on deck. A small auxiliary thermometer is used for correcting change due to air temperature difference.

187. How accurate are reversing thermometers?   These thermometers are handmade from special aged glass and carefully calibrated. They can measure accurately to one or two hundredths of a degree Celsius (also called centigrade). Although so fragile that oceanographers must hand carry them to survey ships, they can withstand pressures of depths as great as 30,000 feet where the pressure is more than 15,000 pounds per square inch. In recent years electronic temperature measuring instruments are becoming more common, but reversing thermometers are still found on most oceanographic research ships.

188. How are reversing thermometers used? Two or three thermometers are placed in a rack attached to a Nansen bottle, a brass bottle which obtains seawater samples for chemical analysis. The Nansen bottles are attached at intervals to a wire cable and lowered into the sea. A brass weight, called a *messenger,* is allowed to slide down the wire, tripping the bottle nearest the surface. As this bottle is tripped, it releases another messenger to trip the next bottle and the process is continued until all the bottles on the "cast" are tripped. As the bottles are reversed, valves close and seawater is trapped. The thermometers are also reversed, preserving a record of the water temperature at that depth.

189. How is the depth of a reversing thermometer determined? A rough estimate of the depth of the thermometers at time of reversal can be obtained from a metal wheel over which the wire passes. This is, however, far from accurate because there is usually a considerable curve in the wire, caused by drifting of the ship and subsurface currents. To obtain an accurate depth, one thermometer is exposed to the effect of pressure and another is protected from pressure. The unprotected thermometer will have a much higher reading because the pressure forces mercury up the stem. The difference between the two thermometers is a measure of the pressure at that depth and pressure is directly related to depth. By a series of calculations, usually carried out on a shipboard computer, depth can be determined with an error as small as half of 1 percent.

190. What is a bathythermograph? This simple but ingenious instrument measures temperature from the surface to depths of several hundred feet. The temperature sensor is a fifty-foot copper tube filled with xylene and wrapped around the fins of the projectile-shaped instrument. As the xylene expands and contracts it moves a stylus which marks a coated glass slide. At the same time, the effect of water pressure on a bellows causes the slide to move, so that the mark is a trace of temperature against depth. By placing a grid against the slide, the temperature can be read in Fahrenheit and depth in feet.

191. What are the advantages of the bathythermograph? The bathythermograph is an inexpensive instrument which can be used while a ship is moving at speeds up to 12 knots. Unlike reversing thermometers, which obtain temperatures at discrete depths, the bathythermograph obtains a continuous record of temperature with depth.

192. What are the disadvantages of the bathythermograph? The depth of the instrument is limited to about 900 feet. Accuracy is limited to one tenth of a degree F at best, not good enough for most scientific studies. In addition, since each instrument has its own nonlinear grid, data from the slides cannot be processed by automated methods and must be read painstakingly by hand. For these and other reasons, the mechanical bathythermograph is being replaced by the expendable electronic bathythermograph.

193. What is the expendable bathythermograph? This is a streamlined instrument which can be dropped from a ship moving as fast as 30 knots to measure temperature from the surface to about 1,500 feet. The temperature sensor is connected to a shipboard recorder by a fine conducting wire which breaks when the maximum depth is reached.

194. What is the range of temperature in the oceans? Temperature in the open ocean varies from about $-2°$ C (28.4° F) to 30° C (86° F). Surface temperatures in the Persian Gulf exceed 90° F in the summer months, and near shore, in shallow water, temperatures as high as 96.8° F have been reported. Most of the water in the oceans is much more uniform in temperature; 75 percent is in the range of 0° C to 6° C and half is between 1.3° C and 3.8° C.

195. Where are the coldest waters found? The coldest waters are found in the Weddell Sea in the Antarctic. Because cold water is heavier than warm water, it sinks and spreads northward along the bottom.

196. Are the highest surface temperatures found at the equator?   The belt of maximum temperature is located about 5 to 10 degrees north of the equator. This is because the Southern Hemisphere has considerably more water than the Northern Hemisphere and is able to absorb more of the sun's heat without a corresponding rise in temperature.

197. How much does surface temperature change from day to night?   This is dependent on cloud conditions, but the daily variation is seldom more than 1° F, and this change takes place only in a thin surface layer. On a clear day, temperature may occasionally rise as much as 4° F. The daily change in water temperature is small compared to air or land temperature because about five times as much heat is necessary to produce the same temperature change in water as in air.

198. In what latitudes are surface temperatures most variable?   Because water at the equator is always hot and water at the poles is always cold, there is little variation either on a daily or seasonal basis. In the mid latitudes there are distinct changes in seasons which cause seasonal variations in the ocean.

199. Are surface temperatures warming or cooling?   Records of the German Weather Service in Hamburg indicate that from 1900 to 1950 there was a warming trend in the North Atlantic. However, from 1951 to 1963, temperatures decreased about 0.3° C.

200. How does temperature change with depth?   In general the temperature in the ocean decreases rapidly from the surface downward. Typically there are three layers. There is a mixed or isothermal layer at the surface which may be 60 to 600 feet thick. Below the mixed layer is a thin zone called the *thermocline* in which there is a rapid drop in temperature. Below the thermocline temperature decreases more gradually.

201. What causes the mixed layer?   The homogeneous upper layer is heated by the sun and stirred by currents, winds, and tides. In areas of constant winds, such as the trade wind belts, this layer may be 600 feet thick. In areas where there is considerable heating during the day a diurnal thermocline may develop to a depth of 20 or 30 feet. A seasonal thermocline may also develop at depths between 100 and 300 feet.

202. How cold is the water at the bottom of the ocean?   Even in the tropics, water at depths greater than a mile is colder than $3°$ C $(37.4°$ F). Temperatures below $0°$ C occur in only a few places.

203. Where does the cold bottom water in the tropics come from?   Studies made about 170 years ago showed that bottom water was so cold that it could have come only from the polar regions. More recent studies of dissolved oxygen indicate that the cold water, being more dense than warm water, sinks in the polar regions and spreads slowly along the ocean floor toward the equator.

204. What is a thermal front?   A thermal front is a definitive line of temperature discontinuity which marks different water masses and water of different origins. These fronts are similar to weather fronts which usually separate different air masses and can be recognized by marked temperature changes over short distances. In the ocean the bathythermograph instrument, which records temperature with depth in a continuous trace, is used to obtain in quantity the observations needed to determine the location of thermal fronts.

205. Does the heat of the earth's crust influence water temperature?   In deep ocean basins below 2,000 fathoms (12,000 feet) the heat of the earth's crust can raise water temperature slightly. This increase is seldom more than $1°$ F.

206. What effects does sunlight have on the ocean?   Sunlight is the source of energy for temperature change, evaporation, and currents. Sunlight controls the rate of photosynthesis for all marine plants, which are directly or indirectly the source of food for all marine animals. Migration, breeding, and other behavior of marine animals are affected by light. And of course light is needed for vision of marine animals as well as human divers.

207. How much sunlight penetrates the ocean?   More than 60 percent of incoming energy is absorbed in the first meter and more than 80 percent in 10 meters. Much more light is absorbed in coastal or turbid waters. Visible light is transmitted much better than either infrared or ultraviolet, and the wave lengths needed by plants are the very ones that are transmitted deepest.

208. What factors affect light penetration? The turbidity of the water is the greatest factor; that is, the amount of matter in suspension, including sediments and microorganisms. The angle of the sun above the horizon is also important. Penetration is greatest at noon. Weather conditions and wave length of light are other factors. Below 200 meters there is probably little seasonal change in transparency.

209. How far can light penetrate into the ocean? In the clearest ocean water when the sun is directly overhead and atmospheric conditions are ideal, a very small amount of blue-green light would be visible to the human eye at a depth of 800 meters (2,600 feet). Instruments have measured light penetration as deep as 1,000 meters (3,300 feet).

210. Why is visibility sometimes better at depths than at the surface? Sometimes horizontal visibility is better at greater depths because of the higher amounts of suspended materials in surface waters. Italian divers working on the liner *Egypt* southwest of Brest, France, reported that visibility diminished as they went to a depth of 66 feet, then improved. Light faded as they reached the wreck at 396 feet; at that depth, visibility was 6 feet.

211. Why do underwater objects appear larger than they really are to skin divers? Underwater objects will appear to be about 30 percent larger to a skin diver using a flat glass face mask. This enlargement is caused by the difference in light refraction in the water and in the air enclosed by the diver's mask. Divers become accustomed to this condition and learn to compensate for it, but it presents serious problems in underwater photography. To correct for the condition, the portholes through which pictures are taken are made of curved glass. When the curvature is designed so that all light rays strike the glass at right angles, distortion will be minimized.

212. How is transparency measured? The method used in 1804 by the U.S. Navy is still in common use today. In that year it was reported that the United States frigate *President*, while en route to

Tripoli, lowered a white china plate attached to a log line and recorded the depth of visibility as 148 feet off the southern Mediterranean coast of Spain. The Secchi disc, long used for measuring transparency, is a white disc 30 cm (about 11.7 inches) in diameter. It is lowered in the water until it just disappears from sight.

213. Where is the clearest water found? Water in the Sargasso Sea in the south-central part of the North Atlantic Ocean approaches the clarity of distilled water. The greatest recorded depth at which a white Secchi disc disappeared in this area is 217 feet. Pacific waters on the average are clearer than Atlantic and Indian Ocean waters.

214. How are more accurate measurements of transparency made? More accurate data can be obtained by instruments containing a photoelectric cell. This cell may either measure the light penetrating from the surface or it may contain its own light source in order to measure the quantity of light that can be transmitted through a fixed distance in seawater.

215. When was the first undersea picture made? A Frenchman, Louis Boutan, a marine biologist of the late nineteenth century is the father of undersea photography. In 1892 he made the first known underwater photo; it was of a Mediterranean sea crab. After working on various camera designs for eight years, his third and last camera was constructed of a heavy copper and iron box and buoyed by a floating wine cask. He wrote a book on undersea photography relating not only the technology and techniques he developed but also reporting his personal views of the value of making photographic records of the undersea biota for scientific use.

216. Who took the first underwater color pictures? Early in 1926, Dr. William H. Longley, a marine biology professor at Goucher College, Baltimore, sent to the *National Geographic* an article on life in the coral reef. Although color photography was still in an early pioneering phase (plates then in use required 1-second exposures), the magazine editors insisted on securing color pictures to accompany the article. Accordingly Charles Martin, chief of the National Geographic photographic laboratory, and Dr. Longley went to Dry Tortugas to obtain the desired photographs. They realized that any underwater pictures would take a great amount of light (flash-bulbs were not yet invented). To meet the illumination requirements, one pound of magnesium powder was exploded on a raft for every picture taken. This lit the bottom 10 to 15 feet below with the equivalent brilliance of 2,400 flashbulbs discharged simultaneously. That did the job and the two men brought back the first set of color photographs of underwater marine life. The January 1927 issue of *National Geographic* carried the results of their work.

217. Can color photography reproduce true colors under water? In clear shallow water the eye can adjust to colors, but the camera records mostly blues and greens. Blue and green filters improve the color considerably, but an artificial light source close to the subject is the best method of obtaining true colors.

218. Where and when were the first undersea movies made? Undersea motion pictures were first made in 1914 by the Englishman J. E. Williamson. The pictures were taken from a specially constructed steel sphere which had plate-glass windows and was suspended 30 feet below a barge; it was large enough to accommodate two men and their cameras. These first movies showed the underwater world of the Bahama coral reefs.

219. How are photographs of the bottom made? Bottom photographs require a source of artificial light. Dr. Harold Edgerton of MIT has been the pioneer in this field. His stroboscopic light makes it possible for a moving ship to take bottom photographs in deep water.

220. How is television used under water? Since 1951 television has been used for locating objects on the bottom, inspecting cables, and studying marine biology and geology.

221. How does the color of organisms vary with depth? Fish living near the surface have a natural camouflage. They are usually dark and tinged with blue or green on top and are silvery or light-colored underneath. At depths where there is still visible light, fish are silvery or pale shades of brown or gray. In the depths of the ocean, marine life is generally dark in color.

222. Why do many deep-sea creatures produce light? Lights appear to be used to provide illumination, to lure prey, and as a signal to attract mates. They may also be used to repel attackers.

223. Why do deep-sea animals have eyes when light does not penetrate to their depth? Some fish migrate vertically to the lighted zone in search of food; others remain at all times in the dark. This second group can see only the light produced by other organisms.

224. Why is the ocean blue? The sea is blue for the same reason the sky is blue, because of the molecular scattering of sunlight. Blue light, being of short wave length, is scattered more effectively than light of longer wave lengths.

225. What causes variations in the ocean's color? Variations in color may be caused by particles suspended in the water, water depth, cloud cover, and other factors. Water in the open sea is commonly blue, especially in tropical or subtropical regions, but may be green near coasts because of yellow pigments mixed with blue water. Heavy concentrations of dissolved material may cause a yellowish hue. Brown color is often caused by mud in suspension. Greenish color is caused by algae in the water. Very heavy populations of minute plants and animals may discolor the water red or brown. The color of the sea changes constantly because of clouds passing across the face of the sun or because of the angle of the sun's rays passing through the atmosphere.

226. What is the Forel scale? It is a basic scale of yellows, greens, and blues used for determining the color of seawater as seen against the white background of a Secchi disc. The U.S. Naval Oceanographic Office uses this descriptive color code scale:

Descriptive color	Code
Deep blue	00
Blue	10
Greenish blue	20
Bluish green	30
Green	40

Light green	50
Yellowish green.	60
Yellow green	70
Green yellow	80
Greenish yellow.	90
Yellow.	99

Water colors are determined by comparing the sea color with a set of standard graduated colors made up of various proportions of ammoniacal copper sulfate and neutral potassium chromate which are contained in a set of 11 vials.

227. What is the density of seawater? Density of seawater depends upon temperature, salinity, and pressure. The average density is 1.025. This means that seawater is 1.025 times the weight of an equivalent volume of distilled water. Increased density results from lower temperature, higher salinity, and greater pressure (depth).

228. How much does pressure increase with depth? Every 33 feet or 10 meters of depth in the ocean increases the pressure by one atmosphere or 14.7 pounds per square inch. By dividing the depth in feet by 2 an approximation of the pressure in pounds per square inch may be obtained. At a depth of 3,000 feet the pressure is 1,450 pounds per square inch, sufficient to squeeze a block of wood to half its volume so that it will sink.

229. What is the pressure at the bottom of the ocean? The pressure on the *Trieste* when at the bottom of the Marianas Trench (35,800 feet) reached 16,883.2 pounds per square inch. The pressure hull which is 3½ inches thick and 7 feet 2 inches in diameter was compressed 2 millimeters by the water over her. Mr. Robert C. Toth calculated that the overall effect was as if two and a half aircraft carriers had been placed on the sphere. The pressure actually loosened the paint. At the bottom of the shallower Puerto Rico Trench (the deepest part of the Atlantic) the French bathyscaph *Archimedes* withstood pressures of about 12,000 pounds per square inch at a depth of 27,510 feet.

230. How is density measured? In nearshore waters, where density varies greatly, a hydrometer is sometimes used to determine density directly. However, in the open ocean where high accuracy is desired, density is not measured, but is computed from temperature, salinity, and depth (pressure).

231. Where is the densest water in the oceans? The water around Antarctica is the densest in the world because it is not only cold but also highly saline. As ice forms, the remaining seawater becomes more salty and more dense.

232. What is Sigma-t? This is the oceanographer's method of expressing density. A density (specific gravity) of 1.025 would be written as 25.0.

233. How does information about density help the oceanographer? Direction and speed of currents can be calculated if the vertical density structure of the water is known. Vertical density structure is also useful in determining whether conditions were stable at the time the measurements were made. If there is an instability (dense water overlying lighter water), mixing will occur.

234. Is seawater compressible? Seawater is almost incompressible. In scientific terms its coefficient of compressibility is only .000046 per bar, under standard conditions. Variations in temperature and salinity will change this value only slightly. The effect of compression is to force the molecules of the substance closer together, causing it to become more dense. If the compressibility of seawater were actually zero, the U.S. Naval Oceanographic Office has calculated that the sea level would actually be 90 feet higher than it is now.

235. How fast does sound travel in the ocean? Sound travels about 4½ times as fast in seawater as in air. Its speed is affected by temperature, salinity, and pressure; an increase in any of these results in an increase in the speed of sound.

236. How is sound velocity measured? Sound velocity can be computed from temperature, salinity, and depth, the three basic measurements made by oceanographers on an oceanographic station. For many years this was the only method used. In recent years sound velocimeters have been used to measure the speed of sound in seawater directly. A sound signal is transmitted for a fixed distance and the travel time recorded.

237. How far can sound travel underwater?   The sounds from depth charges fired by the Columbia University research vessel *Vema* in 1960 were picked up at a distance of 12,000 miles. The depth charges were exploded in a deep sound channel off the coast of Australia and the sound reached Bermuda, almost halfway around the world, approximately 144 minutes later.

238. What is a sound channel?   This is the region where sound velocities first decrease to a minimum value with depth and then increase in value as a result of pressure. Sound waves generated within this layer cannot escape because they are refracted back by the waters above and below and sounds can travel thousands of miles while trapped in this channel.

239. What is SOFAR?   SOFAR (Sound Fixing and Ranging) makes use of the sound channel that is found at depths between 2,000 and 4,000 feet. By triangulation from several listening stations, sound sources in this channel can be located to within about one mile. During World War II many fliers forced down at sea were located by this method. Their planes carried a small bomb which exploded from water pressure when they reached the sound channel.

240. What is sonar?   The word *sonar* was coined from the words *so*und *n*avigation *a*nd *r*anging. Sonar operates on the same principle as radar, but transmits sound waves instead of radio waves. Sonar may be either active or passive. In an active system, a sound is transmitted and the echo received. Distance is computed as one half of elapsed time multiplied by speed of sound in seawater. A passive system is a listening system, and only direction can be determined.

Sonar is used for submarine detection, navigation, fish finding, and depth determination. The depth-finding sonar is commonly called a fathometer, but the correct general name for a depth-finding sonar is *echo sounder*. The word *Fathometer* is a registered trademark of the Raytheon Company and should be used to describe electronic sounders made by Raytheon only.

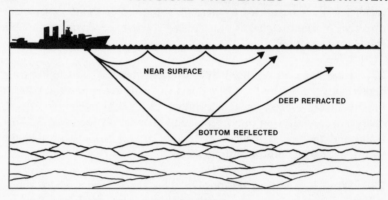

Sound paths

241. What are refraction and reflection?   Sound energy does not travel in straight lines in the ocean because of density differences. It is refracted, or bent, by variations in sound speed of the water; scattered by suspended material or marine organisms; reflected and scattered by the surface and bottom; and attenuated by the water through which it travels.

242. What does ambient noise in the sea refer to?   In the ocean, ambient noise refers to any noise or sound produced in the sea by the environment or by the creatures inhabiting the sea. It includes wave and surf noise, precipitation, seismic and volcanic disturbances, and noises created by fish and all other categories of marine animals. Noise caused by ship, industrial, and other underwater or surface activities which are outside of the platform and measuring equipment itself are categorized as ambient noise also.

VI. WAVES—TIDES—CURRENTS

243. Where do waves come from? The commonly seen waves on the surface are caused principally by wind. However, submarine earthquakes, volcanic eruptions, and tides also cause waves.

244. How are wind waves formed? When a breeze less than 2 knots (2 nautical miles per hour) blows over a calm sea it forms small ripples or capillary waves. As the wind speed increases, larger more visible gravity waves are formed. When the wind reaches a speed of about 13 knots, whitecaps begin to form.

245. With the same wind speed, do more whitecaps appear on oceanic water than on lake water? Research conducted by E. C. Monahan at Woods Hole Oceanographic Institution has shown that the presence of salt in water does produce more whitecaps. Whitecaps consist of many small air bubbles produced by breaking waves. Dr. Monahan's studies have substantiated the belief that more small bubbles are produced in salt water than in fresh. Furthermore, results of his work indicate that these small air bubbles persist longer in salt water. This can be easily observed in a home experiment by simply pouring fresh water from one glass into another, then repeating the procedure after adding table salt to the water.

246. What determines the maximum height of waves? Maximum height depends on wind speed, wind duration, and fetch. Fetch is the distance of open water over which the wind can blow. Wave height in feet is usually not more than half the wind speed in miles per hour, although individual waves may be higher.

247. How does fetch affect maximum wave height? In general, the longer the distance the wind has traveled the higher the waves will be. However, when distances exceed 1,000 miles the waves do not continue to grow appreciably. Maximum wave height under storm winds can be predicted by the formula:

$$H = 1.5 \sqrt{F}$$
where H is the wave height in feet
and F is the fetch in nautical miles

248. Do waves move water over long distances? Waves appear to be moving progressively in one direction, sometimes at considerable speed. However, the water in waves is traveling in a circular motion. Although the wave form advances, the water particle moves forward very little. This can be readily demonstrated by observing an object floating on the surface.

249. Why do waves break? The base of the wave is retarded by friction with the underlying water. The crest of the wave moves faster, so that it curls over in the direction of travel and finally breaks.

250. Why are breaking waves usually parallel to a beach? Waves may approach a beach at any angle, but as they reach shallow water the end nearest the beach slows down because of friction and the wave front swings around to become parallel to the shore.

251. How much power is in a wave? The kinetic energy in waves is tremendous. A 4-foot, 10-second wave striking a coast expends more than 35,000 horsepower per mile of coast. The power of waves can best be visualized by viewing the damage they cause. On the coast of Scotland, a block of cemented stone weighing 1,350 tons was broken loose and moved by waves. Five years later the replacement pier, weighing 2,600 tons, was carried away. Engineers have measured the force of breakers along this coast of Scotland at 6,000 pounds per square foot.

Off the coast of Oregon, the roof of a lighthouse 91 feet above low water was damaged by a rock weighing 135 pounds.

252. Does the energy of the wave alone cause the movement of huge blocks of concrete and other sensational damage at breakwaters? There is a tremendous amount of energy in wave action and the steeper the beach or underwater slope the greater the concentration of wave energy on the sloping surface. Shoals and offshore bars are effective buffers to absorb wave energy before it reaches the coast. However, it must be noted that air and water in crevices of shore structure often act as hydraulic jacks when suddenly struck with the impact of a large wave. Also the air trapped in the pocket created by the collapsing wall of a wave can create tremendous pressures that last only a fraction of a second. Authorities state that such forces which are as high as 14,000 to 17,000 pounds per square foot act with explosive violence.

253. Do waves affect the bottom of the deep sea? The depth at which a wave's effect is felt depends more on wave length than height. Wave motion dies out rapidly and has no effect on the floor of the deep ocean.

254. Why do small waves sometimes cause a ship to roll heavily? When a succession of waves strikes the side of a ship at the same phase of successive rolls, relatively small waves can cause heavy rolling. The effect is similar to that of swinging a child, where the strength of the push is not as important as its timing.

255. Why is it difficult to estimate wave height? Even those with long sea experience find it difficult to estimate the height of waves from a moving ship, because there is no fixed level of reference. It is easy to overestimate the height of a wave when observing it from a ship whose bow is dipping as the wave approaches.

256. What was the highest wave ever observed? On February 7, 1933, the American tanker USS *Ramapo,* proceeding from Manila to San Diego, reported a wave 112 feet high; it was produced by winds of 60–68 knots operating over a fetch of several thousand miles.

257. What are the highest instrument measurements of waves? During storms of hurricane force, waves of 60 to 70 feet have been recorded. One such was a 67-foot wave measured by the British Ocean Weather Ship *Weather Reporter* on September 12, 1961. Such waves are quite exceptional. Waves over 45 feet are not likely to be encountered anywhere except in the center of a hurricane. Waves in the Atlantic rarely exceed 40 feet; in the Pacific they may reach 50 feet. In the stormy Antarctic they may be slightly higher.

258. What is an accelerometer? It is a device which measures the forces of acceleration acting on a body within the instrument. In oceanography it has been used to measure the effect of waves on a ship at sea. Wave patterns and sea roughness can be determined from the instrument's records.

259. Is it the highest waves that cause ship damage? It is the steepness of the waves more than the height that causes the ship to rise at a steep angle or slap with great force. Waves are generally steepest in the early part of a storm.

260. What causes ships to break in two in a storm? A small vessel is sometimes better able to ride out a storm than a large one. Short vessels tend to ride up one side of a wave and down the other; larger vessels tend to ride through the waves on an even keel. If the bow and stern are on successive crests and the center section is in a trough the vessel sags. If the center section arches over a wave the vessel hogs. Under extreme stress it may break in two. This danger can usually be reduced by a change in heading.

261. Can wave heights be predicted? Wave height forecasts were used successfully in World War II for the Normandy invasion; they are now being made routinely by the Naval Oceanographic Office and several commercial organizations. If sufficient information on wind conditions is available, the swell and state of the sea can be predicted a day or more in advance. Wind duration is an important factor, as are wind speed and the distance over which the wind blows.

262. What happens to waves when the wind dies down? When wind ceases, the waves become rounded, less steep, and lower in height. The reduction is so gradual that the waves, now referred to as *swell,* continue until they reach shore; this may be a distance of thousands of miles.

263. What is the difference between sea and swell?

Relatively long waves which have traveled out of the wind-generating area and undergone some decay are known as *swell*. The term *sea* applies to waves still in the generating area. When waves leave the area in which they are formed, the shorter waves die out; the surviving waves exhibit a more regular and longer period with flatter crests. Wave specialists use the term *decay* when referring to this wave degeneration process.

264. What is a surf beat?

Sometimes waves of nearly the same length and originating from different storm areas arrive at a beach simultaneously. When this happens the wave crests from the two different swells may arrive in phase, i.e., coincide with each other, and produce higher waves than either wave train would produce individually. At other times the waves of each wave train are almost completely out of phase, i.e., the wave crest of one wave train arrives about the same time as the trough of the other wave train, thereby having a canceling effect. The increase in water level observed in shallow water caused by the periodic reinforcement and canceling of the waves coming from two different wave groups is known as *surf beat*. It usually has a periodicity of several minutes and is the result of the waves carrying considerable volumes of water shoreward faster than it can escape seaward along the bottom.

265. Why does the sea foam?

Foam is made up of air bubbles separated from each other by a film of liquid. Bubbles coming together in fresh water coalesce, but bubbles coming together in salt water bounce off each other. Most bubbles in the ocean are caused by wind waves, but they may also be produced by rain and even snow. The bubbles that form along the seashore are small, mostly less than ½ millimeter in diameter. When bubbles rise to the surface, they burst and release salt spray into the air. Each bursting bubble causes a jet of several drops to rise to heights up to 1,000 times the bubble diameter. It is believed that most of the airborne salt nuclei come from bursting bubbles.

266. Why is Waikiki Beach a good spot for surfing?

The combination of a nearly flat bottom and the long waves of the Pacific results in waves which steepen until their forward slopes are steeply concave, but do not break prematurely.

267. Where do the "surfing" waves of southern California come from? The swell that breaks on the southern California coast was generated in the equatorial Pacific, thousands of miles to the southwest.

268. Is surfing ever done with equipment other than surfboards? Yes, quite often small boats surf ashore but these are not easy to control and keep at the spot in the breaker where the wave will continue to push them ashore. During some wave studies in 1945 off the Oregon coast, Dr. John Isaacs of the Scripps Institution of Oceanography used an Army DUKW (pronounced "duck") to surf in to the beach on 20-foot breakers. He reported that the DUKW, which operates at 6 knots under full speed, skimmed along at 15 knots ahead of the wave crest.

269. What is body surfing? When started, this sport could be simply described as using the body in place of a board, with the objective of catching a wave, then taking the drop straight off the crest and riding it to the beach. Today, however, the sport has developed with many refinements of technique. For example, more than a dozen riding positions are now used, some for speed and some for stalling. Most individuals participating in body surfing experiment with various maneuvers for the sensation of effortless movement with the water.

270. What are internal waves? These are subsurface waves between layers of different densities. Warmer, less dense water floats on colder, denser water and there is a boundary between, just as there is a boundary between ocean and atmosphere. Because the density difference between water layers is much less than the density difference between water and air, the waves may be hundreds of feet high.

271. What causes internal waves? Little is known about internal waves because of the difficulty in studying them. Tidal phenomena can apparently cause some internal waves; waterspouts, passing ships, or gusts of wind may be other causes.

272. How are internal waves studied?   Towers in shallow water have been used for studying internal waves. In the deep sea, instruments are suspended from ships or buoys. Arrays of buoys with instruments suspended at various depths are the best way of making such studies.

273. What is dead water?   In polar regions or areas off river mouths there is sometimes a layer of fresh water floating on top of salt water. When the thickness of this layer is approximately equal to the draft of a slow-moving ship, internal waves may be produced by the ship's propeller. The energy which would normally propel the vessel is then expended in maintaining these internal waves and the ship becomes sluggish and makes little headway. The phenomenon of *dead water* disappears when speed is increased slightly.

274. What is the phenomenon called cavitation?   Cavitation is the turbulent formation, growth, and collapse of bubbles in a fluid. It occurs when the static pressure at any point in the fluid flow is less than the fluid vapor pressure. Cavitation is generally mechanically induced, as for example by a ship propeller. It generates water noise and causes pitting of propeller blades.

275. What is a tidal wave?   The waves commonly called tidal waves have no connection with tides. The name preferred by scientists is *tsunami,* a word of Japanese origin. They are also called *seismic sea waves* if generated by earthquakes. In addition to submarine earthquakes, landslides or volcanic eruptions cause tsunamis. Most tsunamis originate in the deep trenches along the margins of the Pacific Ocean. Once formed these waves move at speeds that may reach 600 miles an hour. In the open ocean they may have heights of 1 to 3 feet, but as they approach shore they build up to destructive heights.

Tsunamis are described in considerable detail in Barbara Tufty's book *1001 Questions Answered About Natural Land Disasters,* another volume in this series.

276. Can tsunamis be forecast?   Yes, tsunamis can be forecast, because the earthquake waves causing them cross the ocean in only a few minutes and can be picked up by seismograph stations hours before the sea wave arrives.

After the destructive tsunami that struck the Hawaiian Islands in 1946, killing 173 people and destroying 25 million dollars' worth of property, a warning system was set up in the Pacific. Seismograph stations provide information on the time and location of the quake. If the epicenter of the quake is under the sea, a tsunami may result. When a quake is noted, tide stations are alerted to watch for indications of a wave.

A travel time chart centered on the Hawaiian Islands is used to estimate time of arrival of the waves. Warnings of estimated time of arrival are transmitted through an international Pacific-wide communication system. The National Ocean Survey of the National Oceanic and Atmospheric Administration operates the warning service, which has its headquarters in Honolulu.

277. What is the tide? The tide is the continuous cycle of alternately rising and falling sea level observed along coastlines and along bodies of water connecting to the sea. On most coasts, the cycle occupies intervals of about 12 hours, 25 minutes. On some coasts, tides are less frequent. Along some stretches of the Gulf Coast, for example, one tidal cycle occupies about 24 hours and 50 minutes. The rise and fall of sea level observed along coastlines is produced by waves of extreme length; high water is the crest of the wave; low water is the trough.

278. Why observe tides? Observations and measurements of tides are used to determine sea level, the surface from which elevations on land are measured, and to provide engineers with statistical information on reference levels and long-term changes in sea level. Tides can be predicted once observations have been mathematically related to the positions of earth, moon, and sun. Predicted tides are used in oceanographic surveys and as the preliminary reference surface for hydrographic soundings. Engineers use tidal data to determine probable stress limits for bridges, caissons, piers, and other marine structures. Measurement of the currents set up by the same forces which produce tides contributes to safe navigation, to improved harbor design, and to investigations of water pollution and flow. Tide gauges are used in the Seismic Sea-Wave Warning System to detect the presence of potentially destructive seismic sea waves.

279. How does a tide gauge operate?   The gauge is operated by a float that moves up and down with the rise and fall of water in a *stilling well*. The well eliminates the effect of horizontal water movements and, by the size of its intake opening, greatly reduces the effect of rapid changes in water level, such as those produced by wind waves. The up and down movement of the float and attached line operates a worm screw on the gauge and this in turn moves a pencil back and forth across a moving strip of paper. The paper is moved forward at a uniform rate by a clock motor. The combined movement of pencil and paper provides a continuous graph of the rise and fall of the tide. The first automatic tide gauge was installed at Governors Island, New York, in the winter of 1844–45. Recently the addition of electronic telemetry equipment to a standard tide gauge has permitted transmission of tidal records from stations to central recorders.

280. What causes tides?   Gravitational interaction between sun, moon, and earth is the cause of tides. The moon exercises the most influence on our tides; although its mass is much less than the

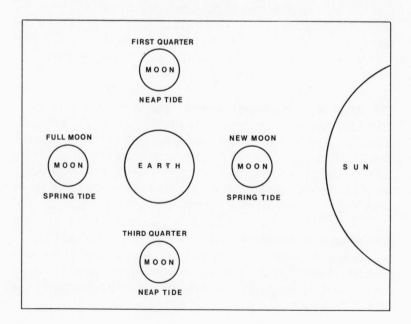

sun's, it is closer to the earth and its tide-producing effect is more than twice as great. Additional effects are produced by land masses, the shape of the ocean floor and water depth, the earth's rotation and the position of the moon relative to the earth. When the sun, moon, and earth are lined up to form new and full moon conditions, we have larger than average ranges in tide, called *spring tides*. When at right angles for the first and third quarter moon conditions, we have smaller than average ranges, called *neap tides*.

The schematic representation shows the positions of the sun and moon relative to the earth at four stages of the monthly tidal cycle. When the attractive forces of the sun and moon are in line, the monthly high or spring tides are produced. When the forces are at right angles, the monthly lows or neap tides result.

281. What is the stationary wave theory?　About 1900 the U.S. Coast and Geodetic Survey developed this theory of tidal phenomena to replace the older progressive wave theory that considered the tide as a single world phenomenon. The stationary wave theory proposes the idea of regional oscillating basins, each with its own natural period and its own responses to tide-producing forces of the sun and moon. The resulting tide in each basin depends on the relation between the natural and imposed periods. Although tidal movement is complicated somewhat by the overlapping of oscillating basins, the theory is consistent with observational data.

282. What is a tidal bench mark?　This is a bench mark set to reference a tide staff; its elevation is determined from the local tidal datum. A tidal bench mark is established wherever tides are observed and the zero mark of the tide staff is tied in to the bench mark by the usual surveying procedures. Elevations of tidal bench marks are determined with reference to the local plane of high water, mean sea level, low water, or other tidal planes.

283. What is high water?　The maximum height reached by a rising tide is called *high water*. Normally, this is due solely to the periodic tidal forces but at times the meteorological effects of severe storms or strong winds may be superimposed on the normal tide to produce high water.

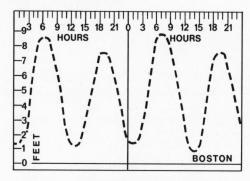

Diurnal type of tide

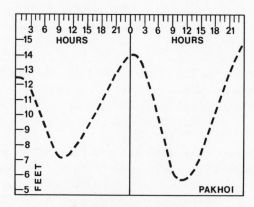

Semidiurnal type of tide

284. What is low water? Low water is the minimum height reached by a falling tide. Usually this is due solely to the influence of periodic tidal forces, but at times meteorological influences of a severe storm or strong winds may be superimposed on the normal tide to produce low water.

285. What are the types of tides? Tides are classified as semidiurnal, diurnal, and mixed. Areas having semidiurnal tides have two high waters and two low waters each day; this is the most common type. Diurnal tides consist of one high and one low water each day. Tides are classified as mixed when they are of the diurnal type on some days and of the semidiurnal type on others. Mixed tides are semidiurnal around the time the moon is on the equator but become diurnal near times of the moon's maximum north or south declination.

286. How long is a tidal day? A tidal day (also called *lunar day*) is the time of rotation of the earth with respect to the moon, or the interval between two successive transits of the moon over the meridian of a place. The *mean tidal day* is approximately 24.84 solar hours in length.

287. How long has the connection between the moon and tides been known? Aristotle in 350 B.C. wrote, "It is even said that many ebbings and risings of the sea always come round with the moon and on certain fixed days." Shortly after the beginning of the Christian era the Roman Plinius made an accurate correlation between the moon and the tides.

288. Who explained the moon's effect on the tides? In 1687 Sir Isaac Newton explained the effect of celestial bodies on ocean water. It was in that year that his law of universal gravitation was published.

289. What factors other than sun and moon affect the level of the sea? Atmospheric conditions, such as winds and pressure changes, can raise or lower sea level. Warm water, being less dense than cold water, will stand higher. River discharge into estuaries or harbors may pile up water to a significant degree.

290. How do hurricanes affect tides? When a hurricane approaches a coast the sea level increases from a few feet to more than 20 feet above the normal tide level. Loss of life from hurricanes is often caused by these rises in sea level. The hurricane of September 1900 in the Galveston area caused the loss of 6,000 lives.

291. Where do the greatest tide ranges occur? The Bay of Fundy in Canada has a mean spring range of 44.6 feet and tides may rise as high as 50 feet. Other areas of great tidal ranges are Bristol Channel and the Sea of Okhotsk.

292. Are there areas of no tides? There are areas with tidal ranges so small that they may be considered tideless. Among these are the Mediterranean, Baltic, and Adriatic Seas and the Gulf of Mexico.

293. How long have tide tables been in use? British Admiralty tide tables giving the predicted height of high water have been published since 1833. A British atlas prepared in 1375 included tide tables and it is probable that such information was available long before that time.

294. How much information is needed to predict tides? Observations at any location must be made over a full 19-year tidal cycle for predictions of maximum accuracy. All significant astronomical modifications will occur within this period. Predictions can be made with considerable accuracy based on one month's observations, as this covers one revolution of the moon around the earth.

295. Who invented the tide predicting machine? The first tide predicting machine used by the U.S. Coast and Geodetic Survey was invented in 1882 by William Ferrel, a mathematician. It was constructed to accommodate constants representing 19 harmonic components of the tide and was used from 1885 to 1911. Rollin A. Horns, another tidal mathematician, together with E. G. Fisher, developed an improved tide prediction machine using 37 components in 1910.

296. Can tides be measured in the open ocean? Yes, it is possible to make deep-ocean measurements with instruments lowered to the sea bottom. These instruments measure the difference in pressure caused by the rise and fall of sea level.

297. What is a tidal bore? Tidal bores result from tidal waters moving up a river against the flow of the river current. As the tidal water enters the river the converging shores cause height and speed to increase. The opposing force of the river current checks this speed and causes the wave to break. This wave of foaming water forms a wall that may be more than 10 feet high.

298. Where is the world's largest tidal bore? The tidal bore in the Tsientang River in China forms a nearly vertical wall of water 8 to 11 feet high from bank to bank of the mile-wide river. Chinese junks sometimes use this wave, moving at 14 miles per hour, to propel them upstream.

299. Can tides be used as a source of power? The French tidal plant on the Rance River is the only one now being used to generate electricity. The success of the French project has caused renewed interest in harnessing the Bay of Fundy. A study is also being made of the Kimberly coast of Australia where tides reach heights of 40 feet. There are several drawbacks to tidal power. Maximum power generation coincides with the tides, not with the demand for power, and in most parts of the world tidal power cannot compete economically with power produced by nuclear fission and other methods.

300. What is the difference between tide and tidal current? The word *tide* is used for the vertical rise and fall of the water, and *current* for the horizontal flow. The tide rises and falls; the tidal current floods and ebbs. The British use the term *tidal stream* for *tidal current*.

301. What are rotary tidal currents? Currents in confined areas such as bays or estuaries flow upstream during flood and downstream during ebb. In the open ocean where currents are not confined to a channel they change direction continually, rather than reversing, and are therefore called *rotary currents*.

302. Do modern ships still sail with the tide? Some of the great liners of today draw up to 40 feet of water. They may make calls at ports with spring low water depths considerably less than this. Scheduling these ships requires a knowledge of time and height of high water and intermediate levels to insure safe clearance.

303. What is a salt-water wedge? Tidal streams or rivers that discharge into the oceans usually carry salt water in their channels. Because seawater is more dense than the fresh water of the stream, it forms an elongated wedge-shaped prism along the bottom, diminishing in thickness as it progresses upstream. With the flood and ebb of the tide, the wedge of salt water penetrates and recedes, respectively. The extent of the wedge also varies with the abundance of fresh-water runoff. When runoff is at a maximum (usually spring) the wedge penetration upstream is least and in periods of low stream discharge (usually late summer or fall) the salt-water penetration is greatest. Cities which use tidal rivers or streams for their water supply have been faced in recent years with a gradual upstream movement of the salt-water wedge because of drought and other conditions contributing to low stream runoff.

304. Do the terms tidewater and tidelands mean the same thing? *Tideland* is land that is covered and uncovered by the daily rise and fall of the tide. Specifically, it is the zone between the mean high-water line and the mean low-water line along the coast. The term *tidewater* is sometimes used synonymously to mean tidelands, but to be technically correct it refers to areas always covered by water. The amount or depth of water (or tide) is immaterial.

305. How do tides affect fish? Some fish feed only when the tides change; others feed only during rising or during falling tides. The grunion, a small fish well known along the California coast, spawns exactly at high tide on the second, third, and fourth nights after the high spring tides.

306. Where do the strongest currents occur? When currents in the open ocean reach 3 knots or more, as for example in the Gulf Stream, they may be considered strong. However, currents in narrow channels, straits, and inshore waters, where tides can create abnormal hydraulic conditions, are much stronger. One of the most formidable currents is in the channels connecting two Norwegian fjords near the Arctic Circle. Three channels connect Salten Fjord and Skjerstad Fjord. In the center channel, known as *storstraum* (which means large current), currents have reached 16 knots during spring tides. When strong westerly and southwesterly winds force more water through the channel into the innermost fjord (Skjerstad), velocities may be even higher. Natives of the area report that when the current flow approaches the maximum, the roar can be heard for miles. These currents create hundreds of whirlpools in the channel, some of which are 30 feet in diameter and several feet deep.

307. How has knowledge of surface currents been obtained? This knowledge is based on millions of observations from merchant vessels. In the 1840s, Lieutenant Matthew Fontaine Maury of the U.S. Navy appealed to shipmasters throughout the world to submit observations of set and drift. From these observations surface current charts were compiled. This procedure is still in use by the Naval Oceanographic office.

308. What are set and drift? The direction toward which a current moves is called *set,* and its speed is called *drift.* When the position of a ship, after a period of steaming, does not agree with the position that should have been obtained by the course and speed of the ship, the discrepancy can be attributed to surface currents.

309. How are drift bottles used? Glass bottles, weighted with sand and containing postcards or return labels, are dropped from ships, ferry boats, aircraft, and even blimps. The finder is asked to note the location and date of discovery. From this information an approximation of drift can be obtained. The Woods Hole Oceanographic Institution releases between 10,000 and 20,000 drift bottles off the east coast of the United States every year. The rate of return has been 10 to 11 percent. Records of all bottles released and recovered are kept in an IBM punchcard system. The data have been used to compile an atlas of surface circulation over the continental shelf.

310. What are seabed drifters? These are plastic cards or envelopes, weighted to drift along the bottom of the ocean. Of 7,000 released by the Woods Hole Oceanographic Institution, about 1,700 were recovered.

311. How long have drift bottles been in use? About 1885 Prince Albert of Monaco used bottles and wooden floats to track currents in the Atlantic. He released about 2,000 floats and received enough returns to prepare a fairly accurate chart of surface currents.

This chart proved useful after World War I in determining probable drift of mines.

312. How far have drift bottles traveled? Woods Hole Oceanographic Institution has records of drift bottles that have crossed the Atlantic from the United States to Ireland, England, and France, a distance of 3,000 miles. Other drift bottles have made a nearly complete circuit, passing the Azores and coming ashore in the West Indies after having drifted 5,000 to 6,000 miles.

Probably the longest undisputed drift on record was a bottle released June 20, 1962, at Perth, Australia, and recovered almost 5 years later near Miami, Florida. Oceanographers at the Tropical Atlantic Biological Laboratory estimated that the bottle had traveled some 16,000 statute miles at a speed of about 0.4 mile per hour. The most probable route was around the Cape of Good Hope, north along the coast of Africa, across the Atlantic to northern Brazil, north along the South American Coast into the Gulf of Mexico, and through the Florida Straits to Miami.

313. How was surface drift first measured in the Arctic Ocean? Fridtjof Nansen, a Norwegian oceanographer, deliberately allowed his ship, *Fram,* to freeze in the ice in 1893 and drifted for three years to prove his theory of a west-moving current. The ice pack covered 1,028 miles at a speed less than a mile per day.

314. How can water masses be traced? Masses of water of common origin differ from each other in salinity and temperature. Movement of water masses can be traced over long distances by their characteristic combination of temperature and salinity.

315. How do oceanographers use oxygen content to trace water masses? As water sinks below the zone of photosynthesis where oxygen is produced, oxygen content is gradually reduced through biological activity. If the water moves over a certain distance slowly, more oxygen will be removed than if the movement had been rapid. Measurements of oxygen content of water over large areas aid oceanographers in tracing limits of currents, but this information must be correlated with other evidence because low oxygen content may be caused by abnormally high biological consumption over a short period rather than normal consumption over a longer period.

316. How is the age of water determined and what does this mean? The most common technique for determining the age of water for different water masses of the ocean is by measuring the decay rate of carbon-14 from a water sample. The half life of carbon-14 is 5,600 years. The age of water is the time that has elapsed since a water mass was last at the surface and in contact with the atmosphere.

317. What is the age of some of the intermediate and bottom water masses? The approximate age, plus or minus 100 years, for the Atlantic Bottom Water is 900 years; for North Atlantic Deep Water, 700 years; for North Atlantic Central Water, 600 years. Antarctic Intermediate and Bottom Water has been measured to be less than 350 years old.

318. What instruments are used to measure currents? Many current meters, both mechanical and electrical, use an impeller to measure speed and a magnetic compass to measure direction. The impeller of the Ekman meter turns a mechanical counter which measures the revolutions during a precise time interval; the impeller of the Roberts meter turns a series of gears to make electrical contacts. In recent years, the Savonius rotor has been used in a number of current meters. Rotations are recorded by an electric counter.

319. What is an Ekman meter? This is a mechanical device invented by the Swedish physicist V. Walfrid Ekman to measure ocean current velocity. A sensitive impeller is turned by the current and the number of turns is recorded on an attached dial. Speed is determined by the number of turns of the impeller; a conversion table supplied with each instrument translates turns into current speed. The meter indicates direction by use of a supply of lead shot which drop into a compass box for a set number of impeller revolutions. From the location of the shot in the compass box, the oceanographer can determine the direction of current flow. The meter is constructed so that it can be operated over a prescribed length of time; scientists on the ship start and stop the meter by dropping messengers down the wire.

Research Vessel *Atlantis*, Woods Hole Oceanographic Institution, sailed about one million miles from 1931 to 1966.

Woods Hole Oceanographic Institution

Aerial view of Woods Hole Oceanographic Institution, Cape Cod, Massachusetts. Below: Aerial view of Scripps Institution of Oceanography, La Jolla, California.

Scripps Institution of Oceanography

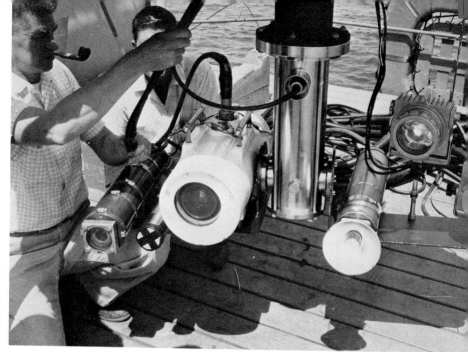

Underwater television camera receives final check. Below: FLIP (Floating Instrument Platform) is towed to area of research and then "flipped" to a vertical position to record oceanographic data.

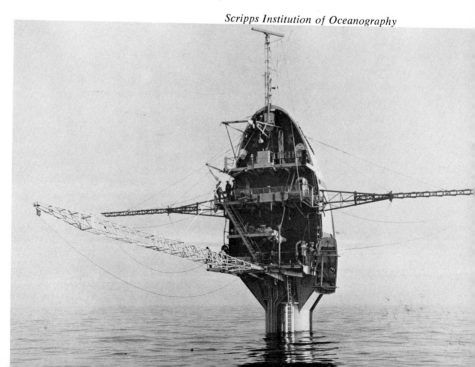

Nansen bottle with reversing thermometers is lowered to predetermined depth to collect seawater sample and record temperature.

National Oceanic and Atmospheric Administration

The STD, an electronic instrument which measures salinity (conductivity),
temperature, and depth (pressure).

Lowering a chain dredge to collect rocks from the ocean bottom.

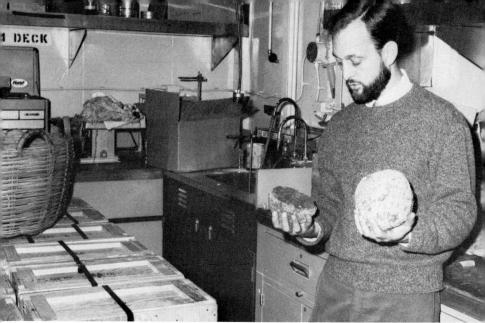

U. S. Naval Oceanographic Office

Scientist examining manganese nodules dredged from the ocean floor.
Below: Marine geologist prepares corer for lowering.

U. S. Naval Oceanographic Office

Forty-foot Coast Guard vessel encounters Oregon surf. Below: Hurricane waves smash seawall.

Icebreaker *Eastwind* rides up on 20-foot-thick ice in Antarctic. Below: Oceanographer measures water temperature in Arctic.

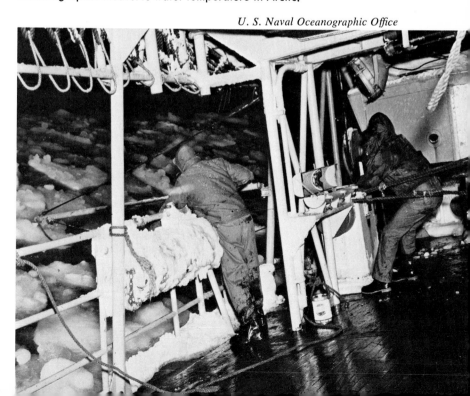

National Science Foundation

Research Vessel *Hero*, a modern oceanographic research ship built for Antarctic waters, has a wooden hull and sails. Below: Iceberg dwarfs Coast Guard Ice Patrol ship *Evergreen*.

U. S. Coast Guard

Waterspouts off Bahamas, May 1961. Below: Aerial view of Surtsey — underwater volcano which surfaced on November 15, 1963. Photo taken two days later.

Robin Clarke

Albino killer whale, named Moby Dick, frolics at Sealand of the Pacific, Victoria, B.C. Below: Trawl taken in Indian Ocean waters by Research Vessel *Anton Bruun* suggests that the area may be a relatively unexploited fishing ground.

National Science Foundation

U. S. Naval Oceanographic Office

Ready to drop the net and begin a plankton tow.

Tropical fish swim past growth of coral at Fort Jefferson National Monument, Florida — America's first Underwater Park. Below: Brain coral, Fort Jefferson National Monument.

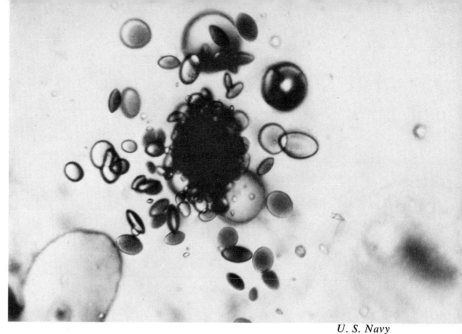

Bacteria with an appetite for oil. A single drop of oil is being devoured by bacteria *(Pseudomonas)*. Below: Crown of Thorns — the coral-destroying starfish.

Grumman's deep submersible *Ben Franklin*. Below: *Trieste* being readied for dive to search for USS *Thresher*.

The Ekman meter must be brought to the surface to record the impeller turns and location of the shot in the compass box after each observation; then it is reset for the next observation.

320. What is a messenger? This is a cylindrical metal weight which is attached to the oceanographic wire and slides down to activate the tripping mechanisms on oceanographic equipment such as Nansen bottles and current meters after they have been lowered to the desired depth. Messengers are usually made of brass; they are approximately 3 inches long and 1 inch in diameter.

321. What is a GEK? Understandably, oceanographers use the abbreviation GEK for the *geomagnetic electrokinetograph,* an instrument which measures currents from a moving ship. The GEK is based on the idea originated in 1832 by Faraday that seawater, being an electrolyte, should generate electric current when it flows through the earth's magnetic field. Two electrodes are towed 100 yards apart. Direction of the current is determined from the mathematical analysis of two tows at 90 degrees to each other.

322. What is the Swallow float? This neutrally buoyant float, invented by the British oceanographer John Swallow, can be weighted so that it will stay at any predetermined depth. It transmits signals which can be picked up by a ship which records its speed and direction. Use of this instrument proved that there are currents far below 2,000 meters (6,500 feet).

323. What is a gyral? All of the oceans have a pronounced circular motion, known as a *gyral*. In the Northern Hemisphere the gyrals move clockwise; in the Southern Hemisphere they move counterclockwise. These patterns are partly caused by the earth's rotation, but the major reason is because the trade winds move from east to west on both sides of the equator.

324. Why are deep-sea currents important? The increasing use of the oceans as a dumping ground for radioactive wastes has caused concern that the wastes will some day return to shore. Based on our present knowledge of deep currents, it appears that the amount of time required for currents to move wastes from the deepest waters to shallow water will allow radioactivity to reach a safe level. However, more information is needed before scientists can say with a degree of certainty that it is safe to dispose of increased amounts of atomic waste in the oceans.

325. How is knowledge of deep circulation obtained?   Most deep circulation has been deduced, rather than measured. Density is calculated from temperature and salinity of seawater samples collected at known positions. Oceanographers predict speed and direction of currents from the density distribution and pressure gradients.

326. How do currents affect climate?   Currents stabilize the climate of adjacent land areas, preventing extremes of temperature. The warm water of the Gulf Stream causes Reykjavik, Iceland, to have a higher winter temperature than New York City. Cold currents have less effect on climate. The major effect of the cold California Current is to produce fog over the west coast of the United States.

327. What is the current flow through the Strait of Gibraltar?   Surface currents in the Strait of Gibraltar flow in only one direction, from the Atlantic to the Mediterranean. Because of excessive evaporation in the Mediterranean, the level is lower than the Atlantic and the Atlantic water is really running downhill through the Strait. Beneath the surface the water flows outward into the Atlantic.

328. How have submarines used these currents?   During World War II, German submarines passed through the Strait of Gibraltar undetected by diving shallow when entering the Mediterranean and deeper when leaving it. The currents carried them noiselessly through the Strait.

329. What is *El Niño*?   At intervals of about 12 years a current of warm water moves down the coast of Peru, displacing the normally cold water of the Peru Current. This current is called *El Niño,* the Child, because it arrives near Christmas. The drastic temperature change results in mass mortality to all forms of marine life and in starvation of the guano birds that feed on fish.

330. What is the phenomenon known as Callao painter?
Mariners refer to the catastrophic destruction of marine life which
blackens the paint on ships within the harbor of Callao, Peru, as
Callao painter. Hydrogen sulfide, released during the decomposition
of the dead organisms, is responsible. The chemistry of the change
resulting in the black color is due to the lead in the paint combining
with sulfur from the hydrogen sulfide to form lead sulfide, which is
black. The immediate cause of this phenomenon is the increase in
water temperature when warmer oceanic currents turn inshore.
Organisms accustomed to colder water die because of the temperature change.

331. What is a rip current?   At points along the shore where
waves are high, water piles up and moves along the shore until an
area of lower waves is reached. At this point it may move out from
the shore as a strong surface current known as a *rip current.* Swimmers who become exhausted in a rip current may drown unless they
swim parallel to the shore. Rip currents are sometimes incorrectly
called *rip tides.*

332. What is the Sargasso Sea?   This name is given to the area
of the North Atlantic east and south of the Gulf Stream system and
within the current gyre. The region is characterized by very clear,
warm, deep blue water which supports very little life except for large
masses of *Sargassum* or gulfweed.

333. Are there currents in the doldrums?   Wind and surface
currents are weak in the equatorial zone known as the *doldrums.*
Below the surface, however, there are *countercurrents,* traveling in
the opposite direction from the surface currents at speeds which may
reach 3 knots.

334. What is a knot?   It is the speed of one nautical mile (6,076
feet) per hour. A speed of one knot is equivalent to 1.69 feet per
second. It is not correct to speak of *knots per hour.*

335. What causes oceanic circulation?   The energy source that
is responsible for oceanic circulation is the heat of the sun. Radiation

from the sun generates winds and causes density differences in the sea. Other factors affecting currents are depth of water, bottom topography, location of land, tides, and deflection by the rotation of the earth.

336. What effects do winds have on circulation?   Winds are the major cause of surface currents. The trade winds blow from the east along the equator, driving the water westward. As speed increases, the currents are given a circular motion by the Coriolis force.

337. What is the Coriolis force?   This is not truly a force, but a result of the earth's rotation which causes water to move to the right of the wind's direction in the Northern Hemisphere and to the left in the Southern Hemisphere. The Coriolis effect is greatest at the poles. The Norwegian ship *Fram,* which drifted in the Arctic ice pack from 1893 to 1896, moved as much as 45 degrees to the right of the wind direction because of the Coriolis effect.

338. What is the Ekman spiral?   Based on ice observations made by Fridtjof Nansen on his Arctic drift in the *Fram* in the late nineteenth century, V. Walfrid Ekman, a Swedish physicist, developed the concept now known as the *Ekman spiral* which has become an accepted theory of the operation of ocean currents. This theory states that a wind blowing steadily over an ocean of unlimited depth and extent and of uniform viscosity would cause the surface layer to drift at an angle of 45 degrees to the right of the wind direction in the Northern Hemisphere. Water at successive depths would drift in directions more to the right until at some depth it would move in the direction opposite to the wind. The velocity will decrease with depth throughout the spiral. The depth at which reversal occurs is about 100 meters (330 feet).

339. How long must the wind blow to generate a current?   In general the wind must blow steadily for about 12 hours to establish a wind current, although there are many other variables. The speed of the current will generally be less than 2 percent of the wind speed.

340. How do density differences create currents? Density differences may produce both horizontal and vertical movement of water, causing modifications to the wind-driven surface currents. In areas of warm, low-density waters the sea surface may be 1 or 2 feet higher than the surface of an area of high density 50 miles away. Water tends to flow from the area of low density to areas of high density. Water may become denser when salinity is increased or temperature is decreased. These conditions prevail in the Arctic and Antarctic, and the resulting cold, high saline water sinks and spreads along the bottom.

341. Why don't water masses mix? Water masses of different temperature and salinity do not mix into a uniform mass because wind and waves affect the upper layer to a depth of a hundred feet or so, and the subsurface currents move slowly. This slow subsurface movement permits mixing only along boundaries between water masses.

342. Does the moon affect currents? Tidal forces appear to affect the Gulf Stream, and probably other currents as well. The Gulf Stream reaches its daily maximum speed about three hours after the transit of the moon. The current is generally faster at the time of neap tides, the tides of decreased range which occur about every two weeks. When the moon is over the equator the Gulf Stream is narrower and faster than at maximum northerly or southerly declination.

343. Who discovered the Gulf Stream? In 1513 three ships under the command of Ponce de León were almost lost in the Straits of Florida. The vessels were heading south from the present Cape Kennedy but moved backward because of the effect of the Gulf Stream.

On September 13, 1492, Christopher Columbus narrowly missed entering the Gulf Stream when he noticed a flight of birds and changed his heading from west to west-southwest. Had he not changed course the Gulf Stream would have carried him to a landfall somewhere between Florida and the Carolinas.

344. Who first charted the Gulf Stream? Benjamin Franklin, while he was Postmaster General, became interested in the reason

Benjamin Franklin's map of the Gulf Stream

why mail ships between the colonies and England traveled faster from west to east. From his study of logs and charts of Nantucket whalers, he produced a chart of the Gulf Stream. During his voyages across the Atlantic he measured the temperature of surface water samples collected with an oaken bucket. This information made it possible for ship's captains to determine when they were entering or leaving the Gulf Stream.

345. Is the Gulf Stream a surface current?   Although the speed of the Gulf Stream decreases from top to bottom, it is still significant at a depth of 1,500 meters (almost a mile).

346. Is the Gulf Stream a river in the sea?   The opening words of Matthew Fontaine Maury's book *Physical Geography of the Sea* (1856) are: "There is a river in the ocean. In the severest droughts it never fails, and in the mightiest floods it never overflows. Its banks and its bottoms are of cold water, while its current is of warm. The Gulf of Mexico is its fountain, and its mouth is in the Arctic Seas. It is the Gulf Stream. There is in the world no other such majestic flow of waters. Its current is more rapid than the Mississippi or the Amazon, and its volume more than a thousand times greater."

We now know that the Gulf Stream is not a river, but a system of separate streams whose movements are complex and irregular. Harris B. Stewart, Jr., has compared the strands that make up the system to the streams and curls of smoke rising from a cigarette.

347. What is the source of the Gulf Stream? The Gulf Stream received its name because of the misconception that its source was the Gulf of Mexico. It is now known that water of the Gulf contributes very little to the flow of the Gulf Stream. Two currents, the North and South Equatorial Currents, join to flow through the passages between the Windward Islands into the Caribbean Sea. The resultant current, flowing through the Yucatán Channel, has only one outlet between Florida and Cuba. Off the southern coast of Florida, other currents coming from the northern coast of Puerto Rico and eastward from the Bahamas add to the flow of the Gulf Stream.

348. How far into the Atlantic does the Gulf Stream extend? The Gulf Stream is most clearly discernible between the Straits of Florida and Cape Hatteras. At the Carolina Cape, the flow assumes a more easterly component, and at about 45° W longitude it divides into three branches. A weak irregular current continues to flow east toward the Bay of Biscay; another current branches off to become the Irminger Current, which turns and flows westward south of Iceland. The main branch continues a northeasterly set to become the Norwegian Current which reaches into the Barents Sea. Water which can be identified by scientists as Gulf Stream water often reaches as far as the northern coast of Novaya Zemlya. Waters originating from the Gulf Stream are responsible for Murmansk, USSR, being an ice-free port.

349. How much water does the Gulf Stream transport? The flow of water through the Straits of Florida is about 26 million cubic meters per second. By the time the Stream reaches Chesapeake Bay, the transport has increased to 75–90 million cubic meters per second because of the addition of Sargasso water and deep water. Beyond the Grand Banks, the flow decreases to less than 40 million cubic meters per second, since part of the water turns southward.

350. How is the position of the Gulf Stream mapped? Since 1966 the Naval Oceanographic Office has been distributing a monthly Gulf Stream Summary showing the observed position and physical properties of the Gulf Stream in the North Atlantic. The current is tracked by an airplane with an infrared thermometer which measures radiation from the sea surface, easily detecting the sudden change of temperature between the Gulf Stream and the cooler coastal water.

351. Is there a countercurrent under the Gulf Stream? Henry Stommel of Woods Hole Oceanographic Institution predicted, on the basis of theoretical considerations, that a countercurrent would be found under the Gulf Stream, flowing in the opposite direction. In 1957 he was proved right; observations with Swallow current floats confirmed a countercurrent at depths of 2,000 to 3,000 meters (6,562 to 9,842 feet) along the east coast of the United States.

352. Are there other countercurrents in the Atlantic? A countercurrent running eastward in the equatorial Atlantic was first noted by William G. Metcalf of Woods Hole Oceanographic Institution. While lowering Nansen bottles he noticed a large wire angle in the direction opposite to the surface current. In 1961 the Woods Hole vessel *Chain* confirmed the existence of the countercurrent. Speeds of 2½ knots were measured at a depth of 100 meters (328 feet). In 1963 the University of Miami research vessel *Pillsbury* tracked the current for 1,320 miles to the west coast of Africa.

353. What is the Cromwell Current? This major subsurface current of the Pacific was originally named the *Equatorial Undercurrent*. It was renamed in honor of one of its discoverers, Townsend Cromwell, a Fish and Wildlife Service oceanographer who was killed in an airplane crash in 1958. The current extends along the equator from the Galápagos Islands to the Solomon Islands or beyond; there is evidence that its total length may be 8,000 miles. It is a shallow current, 120 to 250 miles wide, with speeds of 2½ to 3 knots.

VII. SEA ICE

354. What is the freezing point of seawater? Unlike fresh water, seawater has no fixed freezing point; the point at which ice crystals begin to form depends on the salinity. At a salinity of 35 parts per thousand, the approximate average for the oceans, the freezing point is 28.6 degrees F. As ice begins to form, most of the salt remains in the water, lowering the freezing point.

355. Do waves prevent ice formation? Ice begins to form as a mushy mixture of water and ice spicules. This slush reduces the height of waves. When the mixture freezes it forms a thin, plastic layer of ice which moves with the wave motion.

356. What happens when snow falls on the sea? If snow falls on seawater that is near its freezing point the snow floats on the surface and drifts into beds which may become several feet thick. With a drop in temperature to the freezing point of seawater, the snow and water freeze quickly into a soft ice.

357. Can ice form at the bottom before the surface? If there are no freezing nuclei present, water may cool below its freezing point without freezing. If this happens, the sediments on the bottom serve as nuclei for ice formation and freezing begins at the bottom, rather than the top. Such ice is called *anchor ice*.

358. Why does ice float? If water, like most other substances, became denser as it freezes, ice forming on the surface would sink. As it cools, water becomes denser to a point (4° C for fresh water) near the freezing point, but it then begins to expand and become lighter. At the freezing point its volume is about 10 percent greater than at 4° C.

359. Why does water expand when it becomes ice? When water is cooled it becomes more dense because molecular motion decreases. At 4° C water reaches its greatest density. Between 4° C and 0° C the hydrogen-oxygen bonding within the water molecule is restructured and water begins to assume a different crystalline lattice bonding. This bonding pattern is less dense than that of water above 4° C and occupies about 9 percent more volume. This rearrangement of molecular structure is responsible for the expansion of water when it freezes to ice.

360. What would happen if ice sank instead of floating?   If ice sank to the bottom of the ocean and other bodies of water, it would be protected from the sun's rays and would be very slow to thaw. All the bodies of water would be filled with ice from the bottom up and only the surface layer would melt. The earth would be in a permanent glacial age and would be uninhabitable.

361. Why don't ice caps in high latitudes melt in the summer?   Water requires only one calorie of heat to raise one gram one degree C, but it requires 80 calories to change a gram of ice at 0° C into water of the same temperature.

362. How salty is sea ice?   Sea ice is made up of salt-free crystals surrounding small quantities of brine. New ice contains 2 to 10 parts of salt per thousand. If the ice freezes very quickly a greater amount of salt will be trapped.

363. Why is old ice fresher?   Brine trapped in sea ice is heavier than the ice and tends to seep slowly downward. By the time the ice is one year old its melt water is fresh enough to drink. In the last century, melt water was sometimes used to replenish the water supplies of ships.

364. What is pack ice?   Any large area of ice not fast to the shore is termed *pack ice*. When the pack breaks up, the individual pieces are called *ice floes;* navigable passages through the pack are termed *leads*.

365. How old is the pack ice of the Arctic Ocean?   Ice in the American Arctic has been found to move in a great counterclockwise gyral. Ice in this section of the Arctic Ocean is more or less trapped and may drift around the circular pattern many times. Pack ice which forms north of Siberia drifts eastward across the Arctic Basin and enters the North Atlantic between Spitsbergen and Greenland. Russian ice experts calculate the age of ice in the Eurasian portion of the Arctic Ocean to range from 2 to 9 years. Surface melting reduces ice thickness each summer by 2 to 3 feet; when cold temperatures arrive, freezing begins and ice is added to the underside of the floe. At the edge of the pack, near the coast, none of the ice is very old, as it usually disappears completely each summer. Pack ice in the American Arctic is generally considered to be much older than the ice found on the Eurasian side; ages in excess of 10 years would not be considered unusual.

366. What is ice blink?   The reflection of ice on the underside of a low cloud layer is often the first indication of ice in the distance. This white or yellowish gleam in the sky is called *ice blink*.

367. What is the maximum thickness that Arctic ice can attain in one winter?   The thickness of ice at any Arctic location varies from year to year depending on the severity of the winter. The U.S. Army Engineers Cold Regions Research & Engineering Laboratory, Hanover, New Hampshire, cooperatively with the Canadian Department of Transport, collects observations and maintains a file of ice records for the North American Arctic. Based on a review of recent records for about 35 Arctic stations, the maximum ice thickness attained in one winter is about 9 or 10 feet. However, ice that is rafted or ridged by the wind and currents or ice more than one year old may be considerably thicker. The maximum ice thickness in the Russian Arctic is not appreciably different from the American Arctic.

368. How much does maximum ice thickness increase with latitude in the Arctic region?   One would normally expect the colder weather of more northerly Arctic latitudes to produce thicker ice cover, but this does not always happen. For example, Port Harrison, Canada, reported on June 4, 1965, the maximum ice thickness (108 inches) for all reporting stations (more than 35 Canadian and Alaskan locations). Port Harrison is located on the eastern shore of Hudson Bay at about 58° N, 78° W. The following year, the thickest ice for the American Arctic was reported at Eureka, N.W.T., where

102 inches were measured on June 18, 1966. Eureka is located at 80° N, 86° W! That year Port Harrison reported a maximum of only 83 inches.

369. During the passage of the SS *Manhattan* through the Northwest Passage, what thickness of ice was encountered?
The SS *Manhattan* began its transit of the Northwest Passage in the late summer of 1969, arriving at Prudhoe Bay, Alaska, on September 17. At that time of year ice coverage is near the minimum; in some areas thin layers of new ice are beginning to appear on open water. Observers reported encountering two kinds of ice— annual ice (or new ice) which was as much as 6 feet thick and multiyear ice (old ice) floes which were measured to be as much as 11 feet in thickness.

370. What kind of load will ice support?
According to engineering tests performed by the U.S. Army Engineers Cold Regions Research of Engineering Laboratory, Hanover, New Hampshire, 25 inches of good quality fresh-water ice will support an 18-ton load. Using dual wheel loading, 36 tons could be supported by 35 inches of good quality fresh-water ice.

371. How does an icebreaker break the ice and clear a channel?
The term *icebreaking* is actually misleading. The mechanics of the process are better described as *ice displacement*. Ice forward of the icebreaker is shoved aside to accommodate the underwater portion of the vessel. Where ice coverage is too solid to permit displacement to one side, it is necessary to shove the ice over or under adjacent layers. Only ice too large to be readily shoved aside is broken. Breakage is accomplished by the shock of impact or by cleaving action in which the bow rises up and cuts through because of its weight and the leverage applied at the bow section by the buoyancy of the depressed stern. The broken pieces must still be forced into nearby ice-free areas.

The friction of the ice may bring the icebreaker to a halt. Also the cushioning effect of snow can prevent the upraised bow from returning to the water. When this occurs, liquid ballast is transferred between heeling tanks located on opposite sides of the vessel giving the icebreaker a rolling motion to break the snow's sticky grip. Sometimes, however, even this method combined with the sudden application of full-astern power is of little use.

In cutting a channel, two basic methods are used: One called a herringbone pattern has the icebreaker charging ahead at a small angle to port of the basic course, then to starboard, repeating the process as it weaves back and forth across the channel. The other method called the railroad track technique requires two breakers; they make two straightforward, parallel cuts about 3 ship widths apart. The ice between breaks up under this attack, yielding a channel of greater width than the herringbone pattern.

372. How long have icebreakers been in use in Antarctic waters?   The United States first employed icebreakers in the Antarctic Area in 1946 in connection with Operation HIGHJUMP. The ships used in that operation were the *Burton Island* (under Navy operation at that time) and the Coast Guard's *Northwind*. (All U.S. icebreakers are now operated by the Coast Guard.) These ships proved so essential in the support of that operation that at least three and usually four have been assigned to each of the 14 DEEP FREEZE seasons to date (1969–70). Only two other nations have sent icebreakers into the region: Argentina first dispatched the ARA *General San Martín* to Antarctic waters in 1954 and Japan introduced the newly built icebreaker *Fuji* in 1965.

373. Why is an ice plow more efficient than an icebreaker?   It is easier for an ice plow to break ice by lifting it from below than for an icebreaker to break the ice by pushing down from above, because the ice is supported by water. In recent years the Canadians have used the ice plow to clear ice in the St. Lawrence Seaway. An additional advantage is that the ice broken from below will be pushed out of the channel and onto the adjoining ice.

374. How accurately can ice formation, size, and movement be predicted?   Accuracy of ice forecasting depends on the locale, details required, time range of the prediction, and accuracy of the input weather information. Ice formation predictions are based on heat content and salinity of the water mass, currents, and expected heat exchange from water to atmosphere. The required heat, salinity, and current information is obtained by oceanographers aboard icebreakers when the ice coverage is at its annual minimum. From data so obtained, the *ice potential* of the water can be obtained. With a known ice potential and expected air temperature data applied to the basic laws of thermodynamics one can derive the ice formation forecast.

In the far north, long-range predictions of ice formation are accurate within 2 to 4 days. Farther south, however, where the environmental conditions tend to be more variable, formation predictions are accurate within 8 to 12 days.

Size of the ice pack varies relatively little from year to year in the general area. Variations are mostly on the southernmost fringes where shipping must travel; here variations are of critical importance. Predictions of the size of the pack are therefore generally quite accurate, but the prediction of ice in the shipping lanes needs improvement.

The movement of ice in and out of shipping lanes, or leads, depends substantially on the wind; therefore the accuracy of an ice forecast is dependent on a good wind forecast. An accurate 48-hour to 5-day ice forecast is possible because meteorologists can produce reasonably good wind forecasts. For seasonal ice prediction, which must be based in part on the area climatology, the dates for opening or closing of leads on the Labrador coast may be in error by as much as 6 weeks.

375. Are polar regions becoming warmer?   Receding glaciers and breaking up of ice shelves are evidence of a warming trend. Areas off northern Canada and the Siberian coast which were once frozen have become navigable in recent years. Migrations of fish are another indication of warming; cod have moved farther north along the Greenland coast.

376. Is Iceland surrounded by ice?   In spite of Iceland's far northern location, ice in the surrounding waters was rarely seen until about 1964, the first time in almost fifty years. Since then ice has become a problem from January to June, especially for the Icelandic fishing fleet, Iceland's most important industry.

377. How did the ice ages affect sea level?   The great continental glaciers of the Pleistocene lowered the level of the oceans as much as 400 to 500 feet below the present level. The edge of the continents during the ice ages was near the edge of the present continental shelf.

378. Are the ice caps melting?   The warming trend of the past hundred years has caused ice to melt and sea level to rise about 4½ inches. This warming trend may be part of a long cycle, or the trend could reverse and lead to another ice age.

379. What would happen if all the ice melted?   If all the ice in the Antarctic and Greenland should suddenly melt, sea level all over the world could rise 300 feet or more. The possibility that all the ice would melt suddenly is extremely remote. The time span would most likely be measured in thousands of years and the melting would be accompanied by rising of the land masses and lowering of ocean basins as the weight shifted from the land to the sea.

380. Can Scientists be sure that Antarctica is a solid land mass?   Very little ice-free land area does exist in the Antarctic; only 4.5 percent of the continent is not covered by ice and snow. In some places the ice reaches a thickness of almost 3 miles. This great mass of ice and snow makes Antarctica the highest of all continents; its average elevation is about 7,500 feet. Although the ice sheet conceals almost entirely the shape of the land far beneath, scientists use modern techniques of bouncing radio waves off the land under the ice to map the under-ice land profile. In 1968 they proceeded to drill a hole through the ice at Byrd Station and reached *terra firma* at 7,-100 feet. At the South Pole the ice thickness was measured to be 8,850 feet.

Seismic soundings taken on a 1,400-mile trip from Byrd Station (80° S–120° W) to Eights Coast on the Bellingshausen Sea during the 1961 DEEP FREEZE operations indicated that much of the land under the ice between the Ross Sea and the Weddell and Bellingshausen Seas was below sea level. If the ice were melted, this area may indeed be a group of islands.

381. If all the ice in the Antarctic region melted what would Antarctica look like?   Even if all the under-ice terrain were mapped, it would still be difficult to tell what Antarctica would look like. To begin with, some of the land would still be hidden along the coastline because the melting ice would raise the sea level. No one knows exactly how much the melting ice cap would raise the sea level, but 200 to 250 feet is the usual estimate. Additionally, with the tremendous weight of the ice cap removed, the continent surface would rise considerably, but it would not do so evenly. According to some scientists, the Antarctic land mass would rise almost 2,800 feet if the heavy ice cover were removed. Mapping the topgraphy of the earth's surface under the ice cap is accomplished by seismic soundings. Geologists employ a radar technique whereby sound is transmitted through the ice and reflected from the ice-rock interface.

382. How much ice is there on and around Antarctica?   About 95 percent of the world's permanent ice is in the Antarctic; it is calculated to total about 7 million cubic miles. Of Antarctica's total area of 5½ million square miles, only about 250,000 square miles are free of snow and ice at some time during the year.

In comparison, the Greenland Ice Cap might be considered almost insignificant. Only about 2/3 of a million square miles of that island is ice-covered; the maximum thickness of the Greenland Ice Cap, which has been measured near the center of the island, is a little over a mile (6,200 feet) but scientists figure the average ice thickness for Greenland to be approximately 1,000 feet.

In addition to the ice covering the Antarctic land mass there are vast floating ice shelves that extend over adjacent waters. The largest and probably best known is the Ross Ice Shelf which covers an area of the Ross Sea the size of California. During the long winter the surface of the adjacent sea area surrounding Antarctica freezes as far as several hundred miles from the coast.

383. How do icebergs differ from sea ice?   Icebergs are formed from glaciers or shelf ice and are therefore made up of fresh water; sea ice is formed from seawater. The thickness of sea ice is seldom greater than ten feet, while icebergs may be hundreds of feet thick.

384. What part of an iceberg is below water?   Recent observations indicate that the ratio of depth to height is not as great as was generally believed. For the typical pinnacled berg the depth is two or three times the height. Flat tabular bergs generally extend below water about seven times the height above water.

About 90 percent of the bulk of an iceberg is below the surface because the density of ice is roughly 90 percent of the density of water.

385. How large are the largest icebergs? The largest icebergs are found in the antarctic. In 1956 the U.S. Navy icebreaker *Glacier* reported a berg 208 miles long and 60 miles wide, considerably larger than the state of Maryland.

The largest Northern Hemisphere iceberg ever reported was 7 miles long and 3½ miles wide; this berg was sighted in 1882 off Baffin Island.

386. How high is the highest iceberg? Visual estimates are often greater than measured records. Although many icebergs have been estimated to be 1,000 feet above the water, the highest measurement in the Northern Hemisphere was 447 feet.

The largest tabular bergs in the Southern Hemisphere are about 300 feet above the water.

387. What is the difference between icebergs in the Northern and Southern Hemispheres? Icebergs in the Northern Hemisphere are blocks of glaciers which form on land. The Southern Hemisphere bergs come from the shelf ice along the coast of Antarctica and are tabular in shape, in contrast to the irregularly shaped northern bergs.

388. Can radar be used for detection of icebergs? Radar is used for detecting icebergs but is not wholly reliable because icebergs are not good target reflectors. It is reported that the signal reflection intensity of many icebergs is about half that of an equivalent-sized land mass or ship. Growlers and small bergs with sloping sides may not be picked up by the radar until less than a mile away. The large vertical-sided tabular bergs of the southern oceans and large bergs of the North Atlantic make a much better radar target and are detected by radars at good distances, usually 8 to 10 miles or more. It is reported that discrimination of a radar return from growlers and smaller bergs is particularly difficult when the seas become rough and the waves produce a *sea clutter* return on the scope.

For sea ice, radar can be of assistance to those who have experience in interpreting the scope picture. Smooth sea ice supplies little or no return but hummocky and rafted ice does. A lead in the ice field broken by a preceding vessel is clearly discernible on radar.

389. What sensors and techniques other than shipboard radar are used for ice observation and surveillance? Of all the radar systems, the airborne systems employing side-looking techniques seem to provide the best return and coverage for ice observation. Recently developed infrared sensors have been used with good success for ice detection (and discrimination) from aircraft. Photographs have been an excellent source of ice data since the mid-1940s, providing the best resolution of all. Both infrared sensors and cameras can be restricted by cloudy weather condition; lack of light is a problem for cameras but not for infrared sensors. Satellites in a polar orbit are good remote sensors and provide repeated observations, but clouds can be a problem; resolution is limited. Laser profiling is one of the latest techniques being used for ice observation. At present it is used to determine the topography of the ice field. Underwater sonars are used for ice detection by submarines; sonar also can be used by surface ships for detection of the submerged portions of icebergs and growlers.

390. How many icebergs are calved each year? Authorities estimate that 10,000 to 15,000 icebergs are produced each year by approximately 100 glaciers that reach the sea from the Greenland Ice Cap. Most of these appear along the Baffin Bay Coast of Greenland where the bergs drift south with the West Greenland Current and through Davis Strait where the Labrador Current carries them into the North Atlantic shipping lanes. The number of icebergs reaching the open North Atlantic varies considerably from year to year and depends significantly on the sea ice and wind conditions in the Baffin Bay area. No estimates of icebergs in the Antarctic (Southern) Ocean have been made, although this is where the largest and most spectacular tabular icebergs have been reported.

391. Are icebergs found in the Pacific Ocean? Icebergs seldom if ever appear in the North Pacific Ocean. Glaciers do not reach the open ocean coasts of the North American continent, but are located on the bays, inlets, fjords, and the inland waterway of Canada and Alaska. Bergs that develop from these glaciers are small, by comparison with those calved from the Greenland ice field. Bergs from the few glaciers that reach Pacific coastal waters are usually quite small and melt before reaching the open sea of the North Pacific.

392. How far south can icebergs drift? Large remnants of icebergs have occasionally drifted as far south as Bermuda and the Azores. The distance an iceberg can drift depends partly on its original size and whether or not it drifts into a warm current.

The average number of icebergs drifting south of Newfoundland each year is 400, although the number varies from less than a dozen to more than a thousand.

393. What is the International Ice Patrol? After the *Titanic* struck an iceberg in 1912 and sank with the loss of 1,500 lives, the U.S. Navy Hydrographic Office recommended the establishment of an ice patrol in the North Atlantic. The patrol was begun by two Navy cruisers but is now conducted by aircraft and ships of the U.S. Coast Guard. Seventeen nations contribute to the funding of the Patrol. Since the International Ice Patrol was established, not a single life has been lost through collision with icebergs in North Atlantic shipping lanes.

394. How is the Ice Patrol conducted? Ice surveillance begins early in March when Coast Guard aircraft begin flying from Argentia, Newfoundland, and continues through June or July. Despite man's knowledge of icebergs, his best defense against them is still to track their movements and broadcast warnings. Attempts to destroy icebergs by firebombs, gunfire, and chemicals have all met with failure.

In order to understand the forces of nature that influence the drift of icebergs, oceanographers of the Coast Guard make studies of the origin of icebergs, yearly crop and drift patterns, currents, waves, and meteorological factors. The Coast Guard uses a computer aboard the oceanographic vessel *Evergreen* to aid in predicting the speed and course of icebergs drifting in the shipping lanes of the North Atlantic.

395. What is a growler?   This is a chunk of ice with a cross section between half a meter and 10 meters. It may originate from either sea ice or glacier ice. Larger pieces of ice are called *bergy bits*.

396. How large is the Ross ice shelf?   This ice shelf covers more than half of the Ross Sea which forms a deep indentation into Antarctica. The area of the ice is larger than France. At its seaward edge, where icebergs break off, the thickness varies from 2 to 50 meters (6.6 to 164 feet).

397. What is an ice island?   Ice islands are large tabular pieces of ice calved from glacial ice shelves of the Arctic region such as the Ward Hunt Ice Shelf of northern Ellesmere Island or one of the northern Greenland glaciers. They are similar to the tabular bergs found in Antarctic waters; however, ice islands seldom are reported except in the Arctic Ocean where they drift in a circular track with the currents and wind. Two of the largest bergs sighted in the North Atlantic (in 1882 and 1928) are thought to have been ice islands, as these were reported to be 7 and 4 miles long. The term *ice island* applies to tabular bergs in the northern hemisphere only. They are reported to be 100 to 200 feet thick. U.S. scientists have occupied ice islands on several occasions using them as drifting platforms for making polar geophysical observations. Regular meteorological observations have been taken and many oceanographic studies have been conducted from Fletcher's Ice Island (T-3) and ARLIS I and II. It was proposed that a grounded ice island located off Prudhoe Bay, Alaska, might be used as an offshore oil terminal and docking facility to service icebreaker oil transports which are being considered to move the crude oil produced in the Alaskan Arctic area to refineries.

398. When was an ice island first occupied?   In May 1937 four Soviet scientists landed by aircraft on the ice at the North Pole.

The team made oceanographic and meteorological observations from the floating ice station for nine months.

399. When did Americans first occupy an ice island? Fletcher's Ice Island, also called T-3, was observed by aircraft radar in 1950 and was occupied in 1952 for oceanographic and geophysical studies. It drifts clockwise around the Arctic Ocean. In June 1957 Station Alpha was set up on an ice floe approximately ten feet thick. The station was abandoned in November 1958 because of severe fracturing and hummocking. Another ice island, ARLIS II, was manned from May 1961 to May 1965.

400. What has been learned from ice island observations? Weather observations from drifting stations have increased our understanding of the effect of arctic air masses on temperate regions, and have improved the accuracy of weather forecasts. Oceanographic observations have given us a knowledge of the Arctic Ocean which would have been very difficult to obtain from shipboard.

401. How fast can ice drift? Ice has been known to drift as fast as 50 miles per day in the East Greenland Current during periods of exceptionally strong winds. Typically, ice in the Arctic drifts only a few miles a day.

VIII. AIR-SEA INTERACTION

402. What is a hurricane? Dr. Robert H. Simpson director of the National Hurricane Center at Miami, Florida, describes a hurricane in this way: "A hurricane is a gigantic atmospheric heat pump whose intake reaches out hundreds of miles over the tropical ocean and pulls in moist air from the ocean surface toward a low pressure center. As this air converges near the center around a ring known as the eye-wall, it rises, condenses the moisture which it carries, and releases the latent heat which is the fuel that drives the hurricane. The cloud matter and ice crystals rise in a chimney-like structure imbedded in the eye-wall and is spewed out at an elevation of 8 to 10 miles over vast areas of the environment. This exhaust product is what is normally seen by the meteorological satellite looking down from above and usually is found in a comma-like configuration."

Individuals who have experienced and observed hurricanes first-hand describe a hurricane another way: as "a raging inferno of rolling, swirling waters, of shrieking demonic winds, of lashing rain and of darkness, black and absolute."

Hurricanes are tropical cyclones in which the winds reach speeds of 74 miles per hour or more and blow in a large spiral around a relatively calm, extremely low pressure center known as the eye of the hurricane. The eye may be 20 to 30 miles in diameter and relatively cloud-free but beyond the eye, heavy circular bands of clouds and torrential rains extend for 50 to 100 or more miles in all directions. A fully developed hurricane may cover more than 10,000 square miles of ocean surface.

403. Where does the word "Hurricane" come from? According to Dr. Lahmann-Nitsche, the term *hurricane* appears to have its root from a word *hunrakan* which means *storm god* in the language of the Central American Guatemalan Indians.

404. Where do hurricanes originate? Hurricanes that affect the United States east coast are spawned as tropical cyclones over the tropical and subtropical water of the North Atlantic, Caribbean, and

Gulf of Mexico. According to the National Weather Service, hurricanes are also known to form along the west coast of Mexico and Central America, but these storms seldom move up the coast far enough to affect any western states.

405. What is the difference between a hurricane and typhoon? Hurricanes born and nurtured in the tropical central Pacific Ocean are called typhoons. The significant difference between the Pacific typhoon and the Atlantic hurricane is size; typhoons usually grow to much larger dimensions. Wind speeds in typhoons have been reported to have approached 200 miles per hour. Hurricanes of the Indian Ocean are referred to as typhoons or may be called cyclones.

406. Do hurricanes occur during any particular period of the year? The Atlantic hurricane season normally extends from June 1 to November 30, with most storms occurring in August, September, and October. A few rare hurricanes have been reported in May and December.

407. How does water temperature affect hurricanes? According to Irving Perlroth of the National Oceanographic Data Center, Atlantic hurricanes intensify over warm water and weaken over cool water. His conclusions are based on analyzing tracks of hurricanes and correlating their intensity with sea-surface temperature.

408. How many hurricanes are expected each year? On an average, there are six hurricanes per year with a maximum of eleven observed in a single year (1916 and 1950). In only two years, 1907 and 1914, no Atlantic hurricanes were observed. During the 1893, 1950, and 1961 seasons, four hurricanes were observed in progress at the same time. In 1969 there were thirteen tropical cyclones that developed over Atlantic, Caribbean, and Gulf of Mexico waters, ten of which grew to hurricane proportions. The third one of the 1969 hurricane season, *Camille,* has been described by meteorologists of the Hurricane Research Center in Miami as "the most violent ever recorded for the United States mainland." It killed 256 people and caused almost $1.5 billion damage.

THE TEN YEAR LIST OF STORM NAMES

1971	1972	1973	1974	1975
Arlene	Agnes	Alice	Alma	Amy
Beth	Betty	Brenda	Becky	Blanche
Chloe	Carrie	Christine	Carmen	Caroline
Doria	Dawn	Delia	Dolly	Doris
Edith	Edna	Ellen	Elaine	Eloise
Fern	Felice	Fran	Fifi	Faye
Ginger	Gerda	Gilda	Gertrude	Gladys
Heidi	Harriet	Helen	Hester	Hallie
Irene	Ilene	Imogene	Ivy	Ingrid
Janice	Jane	Joy	Justine	Julia
Kristy	Kara	Kate	Kathy	Kitty
Laura	Lucile	Loretta	Linda	Lilly
Margo	Mae	Madge	Marsha	Mabel
Nona	Nadine	Nancy	Nelly	Niki
Orchid	Odette	Ona	Olga	Opal
Portia	Polly	Patsy	Pearl	Peggy
Rachel	Rita	Rose	Roxanne	Ruby
Sandra	Sarah	Sally	Sabrina	Sheila
Terese	Tina	Tam	Thelma	Tilda
Verna	Velma	Vera	Viola	Vicky
Wallis	Wendy	Wilda	Wilma	Winnie

409. Which cause the most damage—hurricane winds or flooding? According to hurricane experts, winds accompanying a hurricane create much damage but it is a well-established fact that drowning is the greatest cause of hurricane deaths. As the storm approaches and moves across the coastline, it pushes and holds the sea against the shore, raising water levels substantially over the normal high tide, sometimes as much as 15 feet and more. The rise may come rapidly and produce flash floods in coastal lowlands; technically this is called a *storm surge*. Sometimes giant wind waves may be produced which also inundate the coastal lowlands, erode beaches, destroy structures and highways, and cut new channels. These waves are often mistakenly called *tidal waves*. Torrential rains always accompany hurricanes and produce sudden flooding; as the hurricane moves inland and winds diminish, heavy rain continues and floods remain a serious threat.

1976	1977	1978	1979	1980
Anna	Anita	Amelia	Angie	Abby
Belle	Babe	Bess	Barbara	Bertha
Candice	Clara	Cora	Cindy	Candy
Dottie	Dorothy	Debra	Dot	Dinah
Emmy	Evelyn	Ella	Eve	Elsie
Frances	Frieda	Flossie	Franny	Felicia
Gloria	Grace	Greta	Gwyn	Georgia
Holly	Hannah	Hope	Hedda	Hedy
Inga	Ida	Irma	Iris	Isabel
Jill	Jodie	Juliet	Judy	June
Kay	Kristina	Kendra	Karen	Kim
Lilias	Lois	Louise	Lana	Lucy
Maria	Mary	Martha	Molly	Millie
Nola	Nora	Noreen	Nita	Nina
Orpha	Odel	Ora	Ophelia	Olive
Pamela	Penny	Paula	Patty	Phyllis
Ruth	Raquel	Rosalie	Roberta	Rosie
Shirley	Sophia	Susan	Sherry	Suzy
Trixie	Trudy	Tanya	Tess	Theda
Vilda	Virginia	Vanessa	Vesta	Violet
Wynne	Willene	Wanda	Wenda	Willette

410. Why are hurricanes named? Feminine names have been used by the National Weather Service to identify tropical cyclones in the Atlantic Ocean, Caribbean Sea, and Gulf of Mexico since 1953. The first written mention of the use of a girl's name for a storm, as by a forecaster when studying weather charts, may have been in the novel *Storm* written by George R. Stewart and published by Random House in 1941. During World War II the practice of identifying storms by names became a widespread practice of Air Force and Navy forecasters who traced storm movements over the wide expanse of the Pacific Ocean. For several hundred years previously many hurricanes striking the West Indies were named after the particular saints' day on which the hurricane occurred. For examples: Hurricane *Santa Ana* struck Puerto Rico on July 26, 1825, and *San Felipe* (the first) and *San Felipe* (the second) hit Puerto Rico on September 13 in 1876 and 1928, respectively.

Experience shows that the use of girls' names in written and spoken communications is shorter, quicker, and less subject to error than the older, latitude-longitude identification method. These are important considerations when exchanging and reporting information between widely scattered stations, bases, ships, and communities. The use of easily remembered names greatly reduces confusion when two or more storms occur at the same time. Hurricane identifiers should be short, clearly pronounced, quickly recognized, and easily remembered. Before adopting the exact procedure in 1960, the Weather Bureau received many suggestions. Some suggested identification by numbers (1-2-3), by English letters (A, B, C), by Greek letters (Alpha, Beta, Gamma), by boys' names (Arthur, Ben, Christopher), by the phonetic alphabet used by the military in World War II (Able, Baker, Charlie), by names of animals (Antelope, Bear, Coyote), by descriptive adjectives (Annoying, Blustery, Churning), etc. The list ranges from little-known mythological characters and historical personalities to well-known people, places, and things. In 1960 a semipermanent list of 4 sets of girls' names was adopted for use; in 1971 this was replaced by a ten-year list. A separate set of names is used each year beginning with the first set. Letters Q, U, X, Y, and Z are excluded due to the scarcity of names. The list of names follows:

411. Are typhoons and tropical storms of the Pacific identified by girls' names in the same manner as Atlantic hurricanes? The same general scheme using girls' names to identify typhoons and hurricanes is used for the Pacific as well as the Atlantic. One set of 4 alphabetical lists is used for hurricanes and typhoons in the Central and Western North Pacific and another set of 4 alphabetical lists in the Eastern North Pacific. In the Central and Western North Pacific, however, unlike the Eastern North Pacific, each entire list is used; the first name used each year is the name following that used in the preceding year. Names are occasionally changed on the basis of experience with their use. The lists follow:

Eastern North Pacific:

1971—Agatha, Bridget, Carlotta, Denise, Eleanor, Francene, Georgette, Hilary, Ilsa, Jewel, Katrina, Lily, Monica, Nanette, Olivia, Priscilla, Ramona, Sharon, Terry, Veronica, Winifred

1972—Annette, Bonny, Celeste, Diana, Estelle, Fernanda, Gwen, Hyacinth, Iva, Joanne, Kathleen, Liza, Madeline, Naomi, Orla, Pauline, Rebecca, Simone, Tara, Valerie, Willa
1973—Ava, Bernice, Claudia, Doreen, Emily, Florence, Glenda, Heather, Irah, Jennifer, Katherine, Lillian, Mona, Natalie, Odessa, Prudence, Roslyn, Sylvia, Tillie, Victoria, Wallie
1974—Aletta, Blanca, Connie, Dolores, Eileen, Francesca, Gretchen, Helga, Ione, Joyce, Kirsten, Lorraine, Maggie, Norma, Orlene, Patricia, Rosalie, Selma, Toni, Vivian, Winona

Central and Western North Pacific:

Agnes, Bess, Carmen, Della, Elaine, Faye, Gloria, Hester, Irma, Judy, Kit, Lola, Mamie, Nina, Ora, Phyllis, Rita, Susan, Tess, Viola, Winnie.
Alice, Betty, Cora, Doris, Elsie, Flossie, Grace, Helen, Ida, June, Kathy, Lorna, Marie, Nancy, Olga, Pamela, Ruby, Sally, Therese, Violet, Wilda.
Anita, Billie, Clara, Dot, Ellen, Fran, Georgia, Hope, Iris, Joan, Kate, Louise, Marge, Nora, Opal, Patsy, Ruth, Sarah, Thelma, Vera, Wanda.
Amy, Babe, Carla, Dinah, Emma, Freda, Gilda, Harriet, Ivy, Jean, Kim, Lucy, Mary, Nadine, Olive, Polly, Rose, Shirley, Trix, Virginia, Wendy.

412. What is fog? Fog is a hydrometeor which consists of a visible collection of minute water droplets suspended in the atmosphere near the earth's surface. According to international definition, fog reduces visibility below one kilometer (0.62 statute mile). Two types of fog often observed over the water are:

Sea fog, a very common type of advection fog that is caused by warm moist air being transported over a cold body of water, and
Steam fog, which is produced when a much colder air mass drifts across relatively warm water; usually the air temperature must be 9° C colder than the water temperature.

413. What combination of environmental conditions causes ice to collect on ship superstructures? There are three factors that cause ice to form and collect on ship superstructures; freezing rain, arctic frost smoke, and freezing spray. Each condition occurs when the air temperature is below the freezing point of the water; the water drops that come in contact with the ship surface must be very near freezing or supercooled. Freezing rain solidifies on contact and covers the ship with fresh-water glaze ice. The accumulation rate is not usually great. Arctic frost smoke, a term used for steam fog, consists of supercooled water droplets, part of which freezes on contact with the ship; the remainder freezes a short time later. This results in an opaque rime ice. Substantial accumulations of rime ice from arctic frost smoke can occur. The greatest ice accretion results when sea spray is kicked up by winds of Beaufort Force 5 or greater and the ship movement. Air temperatures must be below the freezing point of seawater (minus 1.9° C or 28.6° F) and the water temperature must also be near freezing.

414. Is there any danger of ice accumulating to such a degree that it will sink a ship?   The weight added by ice collecting on the ship superstructure can be substantial; icing can be especially hazardous to small ships such as trawlers. The added weight reduces the freeboard and therefore reduces the range of stability of the vessel. The most dangerous condition occurs when ice collects on high masts and rigging and other superstructure creating a large heeling lever and affecting the ship's stability. Accumulation may even become so great that the ship will become top-heavy and capsize. The windage effect of ice on masts and other rigging also makes ship handling more difficult in rough weather. Loss of trawlers because of ice accretion is well known and documented. Evidence at an inquiry in 1956 into the loss of the trawlers *Loretta* (559 tons) and the *Roderigo* (810 tons) suggests that ice accumulating on these ships from freezing spray may have amounted to 50 tons or more over a 24-hour period. This means that over a 24-hour period, ice can accumulate at rates up to 10 percent of the total weight of a small ship! Serious icing conditions are not restricted to the North Atlantic; the Japanese have reported an instance of 60 tons of ice collecting on a 500-ton vessel in 24 hours.

415. What is an offshore wind?   An *offshore wind* is a wind which blows from the land toward the sea. The opposite phenemona, a wind blowing from the sea toward the land is called an *onshore wind*.

416. What is arctic frost smoke?   This is a weather phenomenon which is a special type of steam fog and is known to seamen by several names. Steam fog occurs when the air temperature is at least 9° C colder than the sea temperature; when air temperatures go below 0° C, it is called *arctic frost smoke*. Frost smoke is usually only a few feet thick. When the top of the fog layer is below the observer's eye level, mariners call it *white frost;* if the layer extends above the observer, seamen refer to the phenomenon as *black frost*. The small water droplets in frost smoke are supercooled because air temperatures are below 0° C. When these droplets come in contact with a ship they freeze immediately into an opaque white rime ice. Reports of the menace of black frost were until recently thought to be greatly exaggerated by seamen but the British fishery research vessel *Ernest Holt* working off Bear Island in the late 1950s reported an ice accumulation of 4 inches in 12 hours on deck and 12 inches on the ship side at the rail level over the same period while operating in dense frost smoke. That accretion rate was calculated to be about 2½ tons per hour!

417. What is the lowest sea-level air pressure recorded for a hurricane or typhoon? According to climatologists of the Earth Sciences Laboratories, Natick, Massachusetts, the lowest sea-level air pressure recorded for a hurricane or typhoon was recorded by a dropsonde observation released from a U.S. Air Force reconnaissance aircraft while making observations of Typhoon *Ida* on September 24, 1958. The recorded pressure was 25.90 inches; it was recorded about 600 miles northwest of Guam at 19° N, 135° E. Other extremely low-surface air pressures, measurements of 26.17 inches and 26.04 inches, were recorded by aircraft flying reconnaissance of Typhoon *Nina* in 1953. The lowest air pressure recorded at the surface was taken on August 18, 1927, when the SS *Sapoerea* recorded 26.185 inches at a location 460 miles east of Luzon, Philippine Islands. Excluding tornadoes, 25.90 inches is the world's lowest air pressure at sea level.

418. If the oceans of the world are a moderating influence on the weather, why are many climatic records established at island or coastal locations? There are a number of climatological records of extreme weather recorded for maritime stations on islands and coastal locations. Some of the most notable are:

World's greatest average precipitation—460 inches at Mount Wai-aleale, Kauai, Hawaii.

Several of the world's maximum rainfall records have been taken on La Réunion Island, located at approximately 21°S, 55° 31'E in the Indian Ocean.

The world's greatest 12-hour rainfall, 53 inches on February 28–29, 1964, at Belouve, La Réunion Island.

World's greatest 24-hour rainfall, 74 inches, March 15–16, 1952, and world's greatest 5-day rainfall, 152 inches, March 13–18, 1952, at Cilaos, La Réunion Island.

The coastal location of Bahía Félix, Chile, is reported to be one of the rainiest locations in the world with an average of 325 days per year with rain observed. In the northernmost part of the same country, two adjacent coastal cities, Arica and Iquique, Chile, are among the world's most arid locations.

Arica has the world's lowest average precipitation, .03 inch.

Iquique has had no rain for 14 consecutive years.

The high incidence of fogginess is another characteristic of coastal and island locations. Cape Disappointment, Washington, is reported to have the record for the United States with an average of 2,552 hours of fog per year. However, Willapa, Washington, reported a total of 7,613 hours in one year.

The foregoing extremes show dramatically the interaction effects of the ocean, the atmosphere, and the terrain topography. Island and coastal mountains are primarily responsible for the high rainfall and other precipitation extremes. Such observations are usually recorded on the windward slopes and the extreme arid conditions on the leeward. Fogginess is caused by the cooling of warm moist air but its persistence in certain locations often results from entrapment by geographic and terrain peculiarities.

419. How much solar radiation is absorbed by the water surface of the earth?   There is a great variation in absorption and reflection of solar energy (radiation) by the oceans. The average reflectivity value computed for all the water areas of the world is approximately 35 percent. The solar reflectivity is least (10 percent) and absorption greatest (90 percent) in tropical latitudes where skies are usually cloud-free. In contrast to the good absorption characteristics of tropical waters, the Arctic Zone with almost a year-round ice cover and persistent cloudiness is an area of high solar reflectivity and low absorption. In the polar regions more than 60 percent and perhaps as high as 80 percent of the incoming solar radiation is reflected.

IX. CHEMISTRY OF SEAWATER

420. How much material is dissolved in seawater? Every cubic mile of seawater contains 165 million tons of dissolved solids.

421. What is the law of relative proportion? This principle, also known as *Dittmar's principle,* was discovered in 1884 by William Dittmar who analyzed 77 samples of seawater collected by HMS *Challenger* on her round-the-world cruise. The analyses revealed that, although the total salt content may vary from place to place, the ratios between the more abundant constituents are virtually constant. This constancy enables scientists to measure one principal element and calculate the other components in seawater, thereby determining its salinity.

422. Why is the chemical composition so uniform? There is constant mixing of waters in the oceans, all of which are connected. Even if water in the deep ocean returned to the surface only once in two thousand years, this would be more than a million times during the estimated three billion years of the oceans' existence.

423. Is the proportion of minor elements constant? No, the ratios of less abundant elements such as aluminum, copper, tin, and bismuth are not proportional to the total salt content. Nor is there a relationship of salinity to dissolved gases such as oxygen, carbon dioxide, and nitrogen.

424. How many chemical elements are in the ocean? At least 72 of the 92 elements that occur in nature have been identified in seawater; most are in extremely small amounts. Probably all the earth's naturally occurring elements exist in the sea.

425. What are the major elements in seawater? In order of abundance they are:

Chlorine
Sodium

Magnesium
Sulfur
Calcium
Potassium
Bromine
Carbon
Strontium
Boron

426. How can artificial seawater be prepared? There are several formulas for artificial seawater. The major salts in each consist of NaCl (sodium chloride), $MgCl_2$ (magnesium chloride), $MgSO_4$ (magnesium sulfate) and $CaCl_2$ (calcium chloride). Lyman and Fleming substitute Na_2SO_4 (sodium sulfate) in their formula for artificial seawater to replace the $MgSO_4$ salt listed in earlier formulas. The approximate values of the major constituents are:

	g/kg
NaCl	26.6
$MgCl_2$	2.3
$MgSO_4$	3.3
$CaCl_2$	1.2
KCl	0.7
Other salts	0.3
(Mainly $NaHCO_3$ & NaBr)	
Total salts	34.4

Add water to make 1,000 g

427. How did the various elements get into seawater? Several of the elements could have been products of rock weathering; these include potassium, sodium, calcium, and magnesium. Others probably originated as volcanic gases; these include the chlorides, bromides, and bicarbonates.

428. Why isn't there more silica in seawater? Although rivers transport about 400 million tons of silica to the oceans every year, there is very little in solution in seawater. One explanation is that the silica reacts with bicarbonates to form clay minerals which become part of the sea-floor sediments. However, it may be that diatoms remove most of the silica from the upper layers of the oceans.

429. What gases are dissolved in seawater?   All of the gases found in the atmosphere are also found in the oceans, but not in the same proportions. Nitrogen is the most abundant, but does not enter into biological processes because of its inert nature. Oxygen and carbon dioxide are also present in significant quantities. Argon, helium, and neon are present in very small quantities. Hydrogen sulfide, which is not normally present in the atmosphere, is found where oxygen is lacking.

430. How soluble is carbon monoxide?   This poisonous gas is produced by a number of marine plants and animals. Everett Douglas of Scripps Institution of Oceanography has determined that carbon monoxide is less soluble in seawater than oxygen, but more soluble than hydrogen or nitrogen.

431. What elements are more abundant in the ocean than on land?   Chlorine, bromine, sulfur, and boron. It is significant that these elements are among the more volatile ones, and their abundance lends support to the hypothesis that volcanic action is largely responsible for the present oceans.

432. In measuring the constituents of seawater, what is the smallest quantity of dissolved solids that can be measured by scientists in the laboratory?   According to information reported by Woods Hole Oceanographic Institution in 1968, Professor Dayton Carritt of MIT has developed a method which permits direct analysis of seawater samples for extremely small quantities of free metals. Described in technical terms the technique is "anodic stripping voltometry using a thin film mercury graphite electrode." During cruise 85 of the R/V *Chain,* chemist W. F. Fitzgerald reported being able to measure very minute amounts of zinc, copper, lead, and cadmium in natural water samples. Quantities of these elements in seawater were measured to as low as 1/100th part per billion.

433. Is seawater acid or alkaline?   It is always slightly alkaline because it contains several alkaline earth minerals, notably sodium, calcium, magnesium, and potassium.

434. What is salinity?   Salinity, or "saltiness," is roughly proportional to the amount of dissolved salts in a given volume of seawater. Oceanographers report salinity in parts per thousands, using the symbol ‰. A salinity of 35 ‰ represents 35 pounds of salt per 1,000 pounds of seawater. Salinity includes all the dissolved material in seawater, not just the salt.

435. What is the difference between salinity and chlorinity?   Chlorinity is approximately equal to the amount of chlorine in the water, as measured by the amount of silver required to precipitate it out. Because salinity is difficult to measure directly by chemical means, chlorinity is usually measured. The relationship between salinity and chlorinity is:

$$\text{Salinity (‰)} = 1.8066 \text{ Chlorinity (‰)}$$

436. What makes the ocean salty?   For the most part the salts were derived from weathering and erosion of igneous rocks of the earth's crust. Some of the salts have been dissolved from rocks and sediments below the sea floor. Gaseous and solid material from volcanoes also contribute to the dissolved solids.

437. Were the seas originally fresh water?   For many years it was assumed that the seas began as fresh water. There is still disagreement among scientists, but it is now generally believed that the seas were either slightly salty initially or became so during the period of formation by dissolving salts from the underlying rocks. Through eons of time, vast quantities of dissolved material have been washed into the oceans.

438. How does salt affect seawater?   Salt increases the density and therefore the buoyancy of water. This buoyancy is of considerable importance to marine animals. The salt contains many chemical elements in addition to sodium and chlorine and these are the materials from which plants manufacture living substance which in turn is the food supply for marine animals.

439. What is the composition of salt in the sea?   Sodium chloride accounts for about 85 percent of the total. Other dissolved solids present in significant amounts include magnesium chloride, magnesium sulfate, calcium sulfate, potassium sulfate, calcium carbonate, and magnesium bromide. There are many other substances present in small amounts. Probably every naturally occurring element known on land is also present in the sea.

440. Is table salt obtained from the sea? Wherever climate is favorable salt can be obtained from seawater by solar evaporation in shallow ponds. Although most of the world's salt is obtained from subterranean deposits (which probably came from the sea), salt from the sea is still a thriving industry in France, Italy, Spain, San Francisco Bay, and elsewhere.

441. How salty is the ocean? In the open ocean the salt content is usually between 33 ‰ and 38 ‰; average salinity throughout the world is 35 ‰. The sea is about as salty as a glass of water containing a teaspoon of salt. The salt in the oceans would cover all the continents to a depth of about 500 feet.

442. Is the salinity of the sea increasing? Undoubtedly the seas are becoming saltier; an estimated billion tons of salts enter the sea from rivers every year. However, there has apparently been very little change over a period of hundreds of millions of years. Primitive animals, such as sharks, have blood which is very similar to seawater. More recent forms of life have the same proportions, but in lower concentration. Although much salt is entering the seas, much is also returning to the land. Salt enters the atmosphere when the sea surface is stirred up by winds. This salt spray is blown ashore or the salt crystals may form nuclei for raindrops. Salt has also returned to the land in sediments which have been uplifted from the sea bottom through the geologic ages.

443. What is the range of salinity in the oceans? Areas of precipitation and dilution by rains or melting ice may have salinities below 10 ‰; the Baltic is one such area. Areas of excessively high evaporation, such as the Mediterranean and Red Seas, may have salinities above 40 ‰.

444. Which is the saltiest ocean? The saltiest ocean is the North Atlantic with an average salinity of about 37.9 ‰; the saltiest part is the Sargasso Sea where high evaporation and remoteness from river discharge increase the salinity.

The Red Sea and Persian Gulf are even saltier, with salinities exceeding 42 ‰. There are spots at the bottom of the Red Sea where salinity exceeds 270 ‰, close to the saturation point.

445. Where are low-salinity waters found?

Melting ice lowers the salinity of polar seas and precipitation and runoff lower the salinity of landlocked seas. The Baltic Sea ranges in salinity from about 2 to 15 ‰; the Black Sea is about 18 ‰. Puget Sound has a salinity of about 20 to 27 ‰.

446. Why is salinity low in the Black Sea?

The enormous amount of fresh water from the Danube, Dniester, Dnieper, and Don Rivers lowers the salinity to about 18 ‰; the average salinity of the open oceans is 35 ‰.

447. How salty is brackish water?

The range of salinity for water described as *brackish* is from approximately 0.50 to 17.00 parts per thousand.

448. Are highest salinities found along the equator?

High salinity is usually associated with high temperature and high evaporation. Along the equator, however, cloud layers may reduce evaporation and frequent torrential rains lower the salinity, so that it is no higher than the average for the world's oceans (35 ‰). The highest salinities coincide with the Tropics of Cancer and Capricorn.

449. How is salinity measured?

Salinity is not measured directly, but is calculated from either laboratory titrations or measurements of electrical conductivity. In the past the most widely used method was to determine chlorinity by titration and then calculate salinity. More recently, electrical conductivity has begun to replace the older method.

450. How is salinity determined by titration?

Silver nitrate is added to a sample of seawater, precipitating silver chloride. The amount of silver nitrate required to precipitate all of the dissolved salts is directly related to chlorinity, which in turn is converted to salinity. The analysis is standardized by comparison with a standard seawater preparation, known as *Copenhagen water,* the chlorinity of which has been determined with extreme accuracy by the Hydrographical Laboratory in Copenhagen.

451. How are salinity samples obtained?   The most commonly used sampling device is the *Nansen bottle,* designed by the Norwegian oceanographer Fridtjof Nansen in the late nineteenth century. This is a brass bottle with valves at both ends. Bottles are clamped to a wire rope and lowered into the ocean. A brass weight, or *messenger,* is then dropped, reversing the top bottle, closing the valves, and releasing another messenger which trips the next bottle. Reversing thermometers are attached to each bottle to measure temperature at the depth of each sample.

A newer type sampler is the *Niskin bottle,* made of plastic in order to collect samples without metallic contamination.

452. How is electrical conductivity measured?   Measurements can be made either in the laboratory or *in situ.* Electrical conductivity depends upon temperature as well as salinity. In the laboratory sample, temperatures are carefully controlled; when conductivity is measured in place, temperature is also recorded. Salinity-Temperature-Depth recorders, called STD's by oceanographers, are lowered to the desired depth and then raised. This results in a continuous trace of salinity and temperature, rather than the individual data points obtained with Nansen bottles.

453. How much oxygen can be dissolved in water?   Dissolved oxygen is measured as milligrams per liter dissolved from a saturated atmosphere at a pressure of 760 millimeters of mercury. The following table gives the amounts for different temperatures of water. It is interesting to note the decrease in the amount of oxygen that water will hold as the temperature increases. The fact that fish, which are cold-blooded animals, require more oxygen as they become more active at higher temperatures, causes the lower saturated oxygen amounts at higher temperatures to be diminished even faster. According to biologists, fish respiration rates double with every 10° C rise in temperature.

Temperature ° C		Dissolved oxygen (Mg/1)
0	–	14.16
5	–	12.37
10	–	10.92
15	–	9.76
20	–	8.84
30	–	7.53

454. How is oxygen lost from the sea? Oxygen may be lost to the atmosphere or it may be consumed by biological activity and decomposition of organic material by bacteria. When abundant oxidizable material is present or stagnant conditions exist, oxygen may become completely exhausted.

455. How has oxygen content changed through geologic time? Until about three billion years ago there was no plant photosynthesis and consequently no excess of free oxygen in either the atmosphere or oceans. As free oxygen became available in the oceans it would have first been used to oxidize various constituents of seawater and would then gradually be dissolved in the seawater.

456. Are there waters in the ocean with no oxygen? Because of the general circulation in the oceans, such areas are rare. The Black Sea, which is almost cut off from the Mediterranean, is totally lacking in oxygen at all depths below 200 meters (656 feet). Below this depth only anaerobic bacteria can exist.

Small basins lacking oxygen have been discovered off the coasts of North Carolina, California, and Venezuela. Some Norwegian fjords having shallow entrance sills are also devoid of oxygen.

X. MARINE BIOLOGY

457. Are any new marine species being discovered? While discovery of new species of marine life does occur, it does not happen these days in the wholesale manner of the nineteenth century when scientists of the *Challenger* Expedition were collecting specimens. This undoubtedly was one of the most fruitful cruises for the marine biologist and expanded many fold the known species of marine life at that time. A total of 4,417 new species were obtained and identified as a result of that expedition.

There is always the possibility that new species may appear in the scientists' nets; more often, however, rare specimens (i.e., only one, two, or very few were previously obtained and identified) are secured these days. Some instances of recently (1968–70) reported rare specimens are:

The Republic of South Africa R/V *David Davies,* trawling at a depth of about 1,300 feet off Natal and Mozambique obtained a rare slit shell, *Pleurotomaria africana.* Slit shells are primitive molluscs usually represented as fossils in early Paleozoic deposits. Until recently they were thought to be extinct; today eight living species are known. There are probably not more than a few dozen specimens in collections.

Another rare find was reported by Mr. Best of the South African Museum. A rare pigmy killer whale, *Feresa attenuata,* was stranded and died at Richards Bay in Zululand. The 10-foot animal was frozen and transported to Durban in almost perfect conditions. Previously, only two skulls were reported in scientific literature. These were finds made in 1827 and 1875. The first description of the external appearance was given in a 1952 account. There is only one other record of *Feresa* from southern Africa; it was a recent stranding of several specimens on the coast of southwestern Africa.

Closer to home is the reported find at Wrightsville Beach, North Carolina, of a rare fish identified as *Acrotus willoughbyi.* According to authorities, only one other specimen was caught previously and that was in 1887 on the Pacific Coast. The fish was caught with hook and line; it was about 5 feet long and weighed about 75 pounds.

The U.S. Naval Oceanographic Office scientist Walter Jahn, in

September 1968, obtained a picture of the rare sea pen, *Umbellulidae,* while working aboard the USNS *Kane.* The photograph was taken about 350 miles off the African Coast in 16,000 feet of water. The animal looks like a flower in bloom but is actually related to jellyfish, sea anemones, and coral. It is attached to the bottom with a rootlike bulb; it has a stalk about 3 feet long with arms and tentacles arranged around a central mouth at the top. Specimens were obtained previously on the *Challenger* and *Galathea* research expeditions; this is the first time it has been seen in its natural habitat.

A notable new species was obtained in 1964 when University of Miami scientists caught a previously unknown specimen near the Bahamas while trawling at about 6,000 feet. It was a black fish about 1 inch long that trails a "lure." Its scientific name is *Kasidoron edom.* More recently (1968) University of Miami scientists aboard the R/V *John Pillsbury* captured a rare fish specimen. While trawling at 26,132 feet in the Puerto Rico Trench they captured a fish of the genus *Bassogigas* which is about 6½ inches long; only three or four other specimens are known to exist in biological collections. It also established a world's record for the depth from which a vertebrate has been taken.

458. What is the difference between plankton, phytoplankton, and zooplankton? *Plankton,* derived from the Greek word for wandering, is used collectively for the community of drifting, floating, or feebly swimming plants and animals in marine and fresh waters. Members of this group range in size from microscopic plants to jellyfishes measuring 6 feet across the bell and include eggs and larvae of fishes and other marine creatures.

Phytoplankton is the collective term for all drifting plant life, from the smallest one-celled plants to large seaweed. Members of this group are the basic synthesizers of nutrients into organic matter by the process of photosynthesis. Phytoplankton makes up an estimated 90 percent of the world's vegetation.

Zooplankton includes the animal forms of plankton; they are the principal consumers of phytoplankton and are in turn the basic food for a large number of fishes, squid, and baleen whales.

459. What are diatoms? They are single-celled algae, mostly less than 0.1 mm long, which are the main food supply for many marine animals. Their siliceous shells form deposits of diatomaceous ooze.

460. What are dinoflagellates?   They are microscopic one-celled plants having characteristics of both plants and animals. Their flagella, or "tails," give them a limited swimming capability. Some varieties have chlorophyll; others do not, and must be classed as animals.

461. What is plankton bloom?   Sometimes plankton multiplies rapidly, producing an enormous concentration. It is usually the phytoplankton which produces a bloom, often a single species. Discoloration of the water may result. The *red tide* is caused by plankton bloom.

462. What causes the red tide?   A red tide, with its mass fish kill, occurs when physical factors are favorable to the rapid reproduction of dinoflagellates (*Gymnodinium*), and the number of predators is temporarily reduced. Dinoflagellates are one-celled organisms with characteristics of both plants and animals. Although less than a thousandth of an inch in size, they reproduce so rapidly that a quart of seawater may contain 100 million. Millions of fish may be killed during such a plankton bloom. More than 50 million were reported off Florida in 1947.

463. Are phytoplankton found in the depths of the ocean?   No, being plants they can grow only within the *euphotic* zone, where photosynthesis occurs. In the open sea this zone extends to a depth of about 100 meters (328 feet).

464. How do zooplankton find food at great depths?   Zooplankton exist on microscopic plants, but below a depth of about 100 meters there is no plant life. Nevertheless, zooplankton have been observed at depths of thousands of feet. The source of their food remained a mystery until the recent discovery that particles of organic matter can be formed by dissolved chemicals in seawater forming particles of proteins, carbohydrates, and fatty acids that cling to bubbles. As the bubbles burst the particles sink.

465. How do plankton remain afloat?   Some plankton have large surface area in relation to weight; others have spines or hairlike projections which increase the frictional resistance to sinking. Planktonic fish eggs contain small drops of oil which increase buoyancy. When the larvae hatch they consume the yolk sac and are buoyed up by the oil.

466. What are the most abundant animals in the sea? Tiny crustaceans known as *copepods* are probably more numerous than all other multicelled animals in the world combined. They are the chief food source for many fish and marine mammals.

467. How do plankton migrate? The migration of plankton communities is vertical, rather than horizontal. The migrations are associated with light. Some zooplankton which are at depths as great as 1,000 meters during the day migrate to the surface at night. They appear to seek a level where light is minimal, but food is available.

Most phytoplankton do not migrate, although some rise to the surface during the day, the opposite of zooplankton migrations.

468. How is water color affected by plankton? Areas where plankton is sparse are deep blue and very transparent. Large concentrations of plankton color the water deep green. Water may be colored yellow or even red by some planktonic organisms. An experienced fisherman can predict the kind and abundance of fish feeding in an area, based on the shade of water color.

469. When was plankton first collected at sea? Perhaps the first scientist to collect plankton was Johannes Peter Müller of the University of Berlin. In 1846 he towed a fine silk net, collecting plants and animals which could be seen only under a microscope. Before this time the existence of the vast numbers of tiny organisms which form the lowest step of the food chain was unknown.

470. What is the half-meter plankton net? This is a conical net held open by a ring half a meter in diameter. It is normally towed for about 30 minutes. Organisms collect in a detachable bucket a few inches in diameter.

471. What is the Clarke-Bumpus sampler? This plankton sampler has a flowmeter to determine the volume of water passing through it during the tow. Its advantage over the half-meter net is that quantitative measurements of plankton per volume of water can be made.

472. What is the Hardy plankton recorder?

Sir Alister Hardy developed this continuous plankton recorder to supplement the data acquired by nets. It consists of an open-ended tube, about 3 feet long, and is designed to be towed at speeds up to 15 knots. The planktonic organisms that enter the tube are picked up by gauze filtering material which is slowly wound across the opening. The winding mechanism is turned by an impeller, so that the movement of the filter is proportional to the towing speed, allowing quantitative as well as qualitative determination of the plankton.

473. Does plankton offer hope for feeding the hungry?

No controlled experiments have been conducted to determine whether man can exist on a diet of plankton. Experiments with rats indicate that they can subsist for extended periods on a diet of plankton and grain, but not on plankton alone.

474. What are some of the problems in using plankton for a human diet?

Planktonic organisms are so widely dispersed that a million gallons of water might have to be filtered to obtain one pound of dry plankton. The price would be many times the cost of the most expensive meat. Some of the organisms which might be filtered from seawater are unpalatable or even poisonous.

475. What is the nekton?

The nekton comprises a large group of animals with the power of location, so that they are able to swim against the current. Nekton includes fishes, whales, adult squids, seals, sea turtles, and many other animals.

476. What is benthic life?

The naturalist Ernest Haeckel proposed the term *benthos* to include all life on or in the bottom of the sea. Some creep, crawl, or burrow; others are attached and are known as *sessile benthos*. Certain fishes, such as flounders, which spend much of their time in close association with the bottom are sometimes considered part of the benthos.

477. How much is known about benthic animals? It is known that benthic invertebrates outnumber free-swimming species by a factor of 10 or 20. But, since most of the benthic organisms occur in restricted areas, they are more difficult to collect and study than the pelagic species. The National Academy of Sciences has cited the lack of knowledge of individual species as a major obstacle to studies on the role of the seabed in replenishing organic matter in the sea and in providing the food of bottom-feeding fishes.

478. What are sedentary species? Sedentary species are organisms that are either immobile on or under the sea floor, or are unable to move except in constant physical contact with the seabed or subsoil. Examples of sedentary organisms include oysters, clams, coral, sponges, and pearl oysters. Shrimp, lobsters, and finny bottom-fish are *not* sedentary organisms.

479. What bottom-dwelling animals are useful to man? Shrimp and oysters are the most important crops in American coastal waters. Other important food animals include clams, scallops, crabs, and lobsters. In other parts of the world, mussels, abalones, conchs, and sea cucumbers are used extensively as food. Among the nonfood items are sponges, pearls, and shells for buttons.

480. What are the most abundant plants in the sea? The diatoms, less than 1/200th of an inch in diameter, are by far the most numerous plants. There may be more than a million in a quart of seawater. Their characteristic feature is a wall of overlapping halves impregnated with silica. They are the most important primary food source of marine animals. Their shells, mixed with calcium carbonate and mineral grains, form deposits of diatomaceous ooze. Diatoms belong to the class of yellow-green algae.

481. What are algae? Algae are primitive plants ranging in size from a single cell, which can only be seen with a microscope, to giant kelps, which grow to a length of 100 feet. Algae are dominant in the sea, both in number of species (approximately 6,600) and in number of individual plants. Although algal cells contain chlorophyll and other pigments, these plants do not have roots, stems, or leaves.

However, some larger forms do have structures which resemble these organs.

482. Why don't algae need roots?   They have no need for roots because they live in a solution of nutrients and the whole plant can absorb both water and nutrients. Some algae have a *holdfast* that resembles a root. The holdfast is simply a structure that holds the plant in place; it does not absorb water or nutrients from the sediments; therefore, it cannot be called a root. Because most of the plant can absorb materials needed for sustenance and growth, there is no need for an elaborate system to transport water, nutrients, and food; therefore algae do not have stems. The supporting structure of kelp that resembles a stem is called a *stipe;* it does not serve a transport function and it does not carry on photosynthesis.

483. Are the leaflike blades of algae used for photosynthesis?   Some algae have blades that resemble leaves, but these are extensions of the plant body and are not the primary site of photosynthesis as in terrestrial plants. Because the entire body of the algal plant carries on photosynthesis, the blades are adapted to increase the surface area to make absorption and photosynthesis more efficient.

484. What is the euphotic zone?   Phytoplankton requires both sunlight and nutrients for growth. The upper layer of the ocean, penetrated by sufficient light for plant growth, is called the *euphotic zone*. Within this zone, photosynthesis is limited primarily by the supply of chemical nutrients. When conditions are favorable, phytoplankton may increase by as much as 300 percent in a day.

485. How deep in the ocean can phytoplankton exist?   In the clearest ocean water, photosynthesis can take place to a depth of about 600 feet; below this depth no plant can exist. In clear coastal waters, sufficient light can penetrate to about 115 feet. In turbid nearshore water, light penetration may be reduced to a few feet.

486. What is the greatest depth for rooted plants?   Rooted plants require more light than floating plants. They are seldom found at depths greater than 65 feet. The greatest depth at which plants have been found growing on the bottom is about 180 feet.

487. Is the seaweed in the Sargasso Sea attached or free-floating? It began its existence attached to the shoal bottom along tropical coasts. Storms uprooted the *Sargassum* and currents carried it to the Sargasso Sea, where it entered the great eddy. The weed continues to grow even after it is uprooted. It is possible that some species of *Sargassum* are permanently free-floating.

488. Is seaweed a weed? Plants as useful as seaweed can hardly be considered as weeds. Seaweed is used as a food by millions of people, as a food for livestock, as fertilizer, medicines, source of iodine, and as an ingredient in the preparation of bread, candy, canned meat, ice cream, jellies, and emulsions.

489. What seaweeds are used as food? During the nineteenth century 75 different kinds of seaweed were used as food in Hawaii, often mixed with seafood. The red alga *Porphyra* is used by the Japanese, who have cultivated it in Tokyo Bay since the seventeenth century. There are presently more than half a million Japanese engaged in processing this alga for use in soups, sauces, macaroni, and other food products.

490. Do seaweeds have high nutritive value? Their nutritive value is comparable to lettuce and other salad materials; they contain a high percentage of indigestible carbohydrates of value chiefly as roughage. The greatest value of seaweeds is their vitamins, including beta-carotene (which is converted to vitamin A by the body), thiamin, riboflavin, panothenic acid, vitamin B-12, vitamin C, and vitamin D. Seaweeds also contain all the minerals essential to normal growth.

491. What medicinal value do seaweeds have? The Chinese and Japanese used seaweeds in the treatment of goiter and other glandular troubles as long ago as 3000 B.C. Various seaweeds have been used in the past to prevent scurvy, as vermifuges, as laxatives, to reduce fevers, and to treat stomach disorders.

Some species of seaweeds have been found to have anticoagulant and antibiotic properties. Investigations are now being conducted in ulcer therapy.

492. What uses have been made of kelp? Products which have been made from kelp include iodine, sodium chloride, potassium chloride, bleaching agents, acetone, ethyl acetate, and other solvents. During World War I, ten American factories were engaged in processing kelp harvested along the California coast. If it ever becomes necessary, these kelp could again be used as a source of potassium chloride for fertilizer. Gelatinous products are also extracted from kelp and Irish moss.

493. How is Irish moss used? Irish moss is harvested off the coast of New England. Extracts of algin and agar-agar are used in ice cream, chocolate milk, pie fillings, soft drinks, dietetic foods, and many others. It is also used in pharmaceuticals and cosmetics.

494. Can marine plants live under ice? University of Alaska scientists have found eelgrass living under winter ice. Eelgrass is a marine plant common to the Pacific coast and is found from the Gulf of California to the Chukchi Sea. The plants were observed through a submarine television system; they were reported in good vegetative condition, growing in about 8 to 12 inches of water between the sediment and the ice. The water odor indicated the presence of hydrogen sulfide and instrumented tests indicated that the water was lacking in oxygen.

495. How large do plants grow in the ocean? The giant Pacific kelp, *Macrocystis pyrifera,* may grow to a length of almost 200 feet. Giant kelps grow in water 10 to 30 meters (32.8 to 98.4 feet) deep. Under the most favorable conditions, the giant kelp can grow as much as 12 inches a day.

496. Why do plants grow better in some areas than in others? Availability of phosphates and nitrates is the major reason for plants growing more luxuriantly in some areas. In addition, iron, copper, and manganese are needed for good growth. Plants thrive best in shallow nearshore areas where light can penetrate. Stirring of nutrients from the depths by upwelling currents also promotes growth.

497. Why are large plants rare in the upper layers of the open sea? As plants multiply they deplete the supply of phosphates and nitrates and only the plants which absorb nutrients most efficiently can survive. Small plants have a large surface relative to their mass. Therefore, one-celled algae are the dominant form in the open sea. They are extremely numerous and are referred to as the "grass of the sea" because they are the very beginning of the food chain in the sea.

498. How does the sea compare to the land in productivity? The most productive areas of the ocean can produce less than one tenth the amount of dry organic matter per acre that fertile farm land can produce. The reason is that rich soil contains thousands of times as much nitrogen and phosphorus as seawater.

499. How do plants vary in color with depth? As sunlight penetrates the sea, light at the red end of the spectrum is absorbed. Attached plants must, therefore, adapt to the available light. Yellow-green algae are found near the surface. As depth increases, plant colors become blue-green, green, brown, and red.

500. Is it possible to bring nutrients to the surface to feed plants? Nutrients are brought to the surface by natural upwelling, and the idea of bringing the nutrients to the surface by artificial means is intriguing. One idea is a source of compressed air on the sea bed to form an upward current; another idea is warming the bottom water by a nuclear source so that it will rise. Neither idea would be practical on a large scale.

501. Will plants from the sea feed future generations? Although algae are high in protein and can be artificially flavored to taste like meat or vegetables, they are mostly microscopic and difficult to harvest economically. The large plants are easier to harvest but they grow in only a small part of the sea. It is probable that the main source of food from the sea will continue to be fish and invertebrates.

502. Are salt-water marshes suitable for grazing use? According to the U.S. Department of Agriculture, the coastal marsh belt from Southern Texas to Florida contains millions of acres of grazing lands. In some locations the marsh is only a few hundred yards wide; in other places such as the Mississippi River Delta, it is more than 50 miles wide. The inland portion of this belt is fresh-water marsh; the salt-water marsh covers a broad zone extending from the fresh-water conditions to the coast. Families originally settling the marsh attempted to cultivate the higher ridges; those living in the marsh area today have turned more and more to using this land for grazing of cattle and hogs for their livelihood. Hazards to cattle grazing these lands include heat, insects, disease, heavy rainfall and wind tides that may submerge the grazing area, lack of shelters, and unstable soils. Other problems include lack of fencing, inadequate water for livestock to drink, uncontrolled marsh fires, too many undesirable plants, and lack of reserve feed supplies for critical periods. However, the numbers of livestock grazing in salt-marsh lands have increased despite difficult conditions. Good results from planned breeding of a special strain of cattle well suited for being raised in marsh areas have resulted in animals that endure heat, withstand insects, and if necessary graze belly-deep in water.

503. What are some salt-tolerant plants that are suitable forage?   Salt-marsh grasses that are of highest forage value are:

> Marshhay cordgrass (*Spartina patens*)
> Big cordgrass (*Spartina cynosuroides*)
> Smooth cordgrass (*Spartina alterniflora*)
> Seashore saltgrass (*Distichlis spicata*)
> Common reed

These grasses provide the best range conditions. Seashore paspalum (*Paspalum vaginatum*) and longtom (*Paspalum lividum*) are crawling grasses that provide good forage and show good increase on overgrazed marsh. Longtom provides good hay production but will tolerate only slight salinities. The hairypod cowpea (*Vigna repens*) is the most common legume in the marsh. It prefers slightly elevated areas, is an annual, and therefore not very dependable for grazing. In good seasons it makes a large volume of forage and is readily eaten by cattle. Salt-marsh range will carry one animal to 4 acres if in excellent conditions; the acreage requirement ranges upward to 12 acres for ranges classified in lesser conditions.

504. How many species of fishes are there?   There is little agreement among scientists on the number of species. Estimates range from 15,000 to 40,000 species; however, 25,000 appears to be the most often quoted figure. This discrepancy exists because fish species are sometimes named more than once because of inadequate descriptions and variations caused by environment or geographic distribution. In some fish species, the male has been described as belonging to one species and the female to another because of a difference in body form or color pattern. This phenomenon is called *sexual dimorphism*. Other fishes have been named more than once because the young look different than the adults. In addition, most scientists agree that not all fishes have yet been named; the estimate of 25,000 allows for this unknown.

505. How do bony fishes compare with cartilaginous fish in number of species?   The species of fishes with bony skeletons are more numerous than those with skeletons of cartilage (sharks and rays). Bony fish number around 20,000 while the cartilaginous fish number only about 600.

506. Are all fish cold-blooded creatures?   Cold-blooded animals maintain the same body temperature as their surrounding environment. Most fishes are cold-blooded. Exceptions are tuna and certain sharks. Recently physiologists have become interested in studying these fishes to learn how they maintain the body-environment temperature differential.

507. Why do some fish lay more eggs than others?   Sharks and skates lay eggs that are enclosed in a protective capsule. Because survival until hatching time is ensured, these fish lay only a few eggs. Other species that spawn in the open sea and produce eggs that float unprotected may lay half a million at one time.

508. How can the age of fish be determined?   Fish scales have rings similar to the growth rings of trees. The number of rings is related to age and the width indicates the growth rate in one season.

509. What is the largest bony fish?   The heaviest is probably the sturgeon. Although the sturgeons are largely restricted to fresh water, some of the largest spend most of their lives in the sea. Specimens more than 25 feet in length and weighing between 2,500 and 3,000 pounds have been caught in Russian rivers. Other fish that can exceed a ton are the blue marlin, black marlin, and bluefin tuna.

510. How widely do tuna travel? Bluefin tuna tagged and released off California have been recaptured several years later off Japan.

511. Do tuna ever stop swimming? Tuna are heavier than salt water; if they were to stop swimming they would sink tail-first to the bottom where they would suffocate because of the lack of oxygen.

512. What fishes spend part of their lives in fresh water and part in salt water? The term *anadromous* is applied to fishes that ascend rivers from the sea to breed; salmon and shad are examples. Salmon spend most of their lives in the ocean, but return to inland streams or pools to breed. The life cycle of the eel is reversed. It begins life in the mid-Atlantic and then migrates to Europe or the United States. It remains in fresh water as long as 15 years before returning to the mid-Atlantic to spawn and die.

513. How does the salmon navigate for thousands of miles to find its spawning ground? How salmon migrate great distances and unerringly find the stream of their birth has long intrigued scientists and is still unknown. Their method of finding the right tributary is better known. Experimental evidence indicates that they are guided by the characteristic odor of the bottom sediments.

514. Is life found at all depths in the ocean? This question was settled for all time when Piccard and Walsh reported a flatfish, resembling a sole, at a depth of 35,800 feet. From the porthole of the bathyscaph *Trieste* they observed a fish about 1 foot long and 6 inches wide swimming away. As recently as 1860 some scientists believed that marine life could not exist below 1,800 feet. This view was discredited when a telegraph cable brought up from a depth of 6,000 feet was found to be covered with a variety of marine life.

In 1872, scientists aboard HMS *Challenger* found life in the deepest areas which they were able to trawl, but it was not until steam trawls and wire rope became available that trawl collections could be obtained from the deepest trenches. In 1951 the Danish oceanographic ship *Galathea* dredged various invertebrates from a depth of 33,433 feet in the Philippine Trench and a year later caught fish at a depth of 23,400 feet.

515. How do fish at great depths differ from those at the surface? Swim bladders may be reduced in size or be absent altogether; buoyancy is provided by fat rather than gas. Many deep-sea fishes have very large mouths and are capable of swallowing fish larger than themselves.

516. Do fish at great depths have eyes? The eyes of some fishes living at depths less than 6,000 feet are very large and efficient. At 6,000 feet there is no light, but the fish sometimes come to within 1,000 feet of the surface where their eyes may help them find food. Below 6,000 feet the eyes are often small or degenerate or may be absent altogether. There are some exceptions; some species living on the bottom at depths of 15,000 feet have large eyes. In this region eyes can only detect the light of naturally luminescent organisms.

517. Why are only female specimens of the angler fish found in the sea? This question perplexed scientists who captured this small but fierce-looking fish from the dark ocean depths. After prolonged investigations, marine biologists discovered that a male was captured along with each female; the male is physically attached, but the attachment is not obvious in mature specimens. Early in the life cycle, the male finds his mate and promptly attaches himself by sinking his teeth into her side. He remains there for life, obtaining nourishment directly through this attachment. Most of his vital organs gradually cease to function; only the reproductive glands continue to grow to maturity. Gradually the male becomes completely enveloped by flesh as the female grows to maturity, and finally he appears as little more than a bump on her side.

518. Do fish make sounds? Among the noises produced by fish are croaks, grunts, coughs, whistles, and squeaks. Fish also make grinding, drumming, and rasping sounds. Not all fish make sounds; some, such as the flounder, are practically mute.

519. How do fish produce sounds? Most fish do not have vocal cords. The most usual ways of producing noise are by vibrating their swim bladder or rubbing parts of their skeleton together. Some species snap their fins or gnash their teeth.

520. Why do fish make sounds? The drumlike thumps of drum-fish and groupers are believed to be a defense to frighten other approaching fish. The "boat whistle" sound of toadfish may be related to reproductive behavior. Other sounds may be used to warn the school of approaching danger. Actually, very little is known about the reasons fish make sounds.

521. Do fish have good hearing? Investigations on tuna by the Honolulu Laboratory of the National Marine Fisheries Service show that tuna can neither hear or see as well as humans. The tuna could not hear sounds above 2 kHz; humans can detect sounds to 15 kHz or more, three octaves higher.

522. Do fish have a sense of smell? Salmon find their way through a series of tributaries to the place of their spawning by smell. The eel, which also migrates back to the place of its birth, also has a keen sense of smell. Only a few fishes appear to locate their prey by smell, notably those that live at great depths. For other fishes smell does not serve a major purpose.

523. Do fish have a sixth sense? Fish have sensory systems extending along the flank and onto the head, consisting of cells with a protruding hairlike element. Water currents stimulate the sensors and alert the fish to movement nearby; some fish can detect movement at distances greater than 50 feet.

524. Does a flying fish really fly? This question has been settled by high-speed photography. The flying fish does not vibrate its wings during flight, but merely glides. To leave the water the fish moves upward rapidly and beats the water with its tail. It travels through the air at about 35 miles per hour.

525. Is there actually a fish that walks on the bottom? The fish with the scientific name *Benthosaurus* has 3 fins that have been modified down through the ages and look like three spindly legs on which it stands and moves in "hops." Initially, scientists thought that these 3 long fins served as feelers but observation of the *Benthosaurus* in their native bottom habitat from a bathyscaph in 1957 proved this opinion in error.

526. Are eels and sea lampreys closely related biologically? They are related only in the sense that each is a fish. The sea lamprey, *Petromyzon fluviatilis,* is one of the most primitive fish. The hagfish and lamprey are called cyclostomes, which means *round mouth.* They belong to the class Agnatha, which are jawless fish. Biologically speaking, eels, *Anguilla rostrata* and *Anguilla vulgaris,* are more advanced species having well-developed jaws and teeth. Eels feed on most any animal food, living or dead, but lampreys are parasites which attach to a fish, rasping their horny circular mouth into the flesh and sucking the blood. Both are elongated creatures; each may grow to a length of almost 3 feet. They are not closely related species.

527. Do eels breed in fresh-water streams or in the ocean? Eels have been known to man and used as food for thousands of years, but only recently has their breeding ground been located. In 1856 a German naturalist described a leaf-shaped fish he had found in the sea and called it a *leptocephalus.* Almost a half century later, in 1896, two Italians, Grassi and Calandruccio, noticed the resemblance of the leptocephalus to the eel. After a Danish research ship found a leptocephalus north of England, the first ever found outside of the Mediterranean, the Danish Government offered Johannes Schmidt the job of finding the breeding ground of the eel. This task required 17 years of painstaking research and surveys which crisscrossed the North Atlantic and Mediterranean in every direction. Finally he found leptocephali less than one half inch long. In 1922 he announced the breeding place as the Sargasso Sea, about 1,500 miles off the Florida coast. Here breeding eels descend to depths of perhaps 1,500 feet to release the eggs. It is presumed that the eels die after mating. After hatching, the larvae drift with the currents. The European eels take about 2½ years to travel the 3,000 miles to their coasts. The American eels, which are a distinctly different species, are thought to breed in the same general location and reach the eastern coast estuaries in a few months.

528. Are there any unique differences between the American and European eels? During his study of eels which continued for almost twenty years, the Danish biologist Johannes Schmidt undertook to explain the differences observed in fresh-water eels. They vary in color from olive or muddy brown to black. There are often differences in the shape of the head and eyes and the appearance of the digestive organs, especially in large eels. Factors such as color are determined by the type of bottom on which they live; other factors are dependent on whether or not the eel is getting ready to go back to the sea to breed. During this investigation of the physiology of eels, Schmidt discovered that the American and European eels are different and can always be distinguished by a count of the bones in the spinal column. All European eels (*Anguilla vulgaris*) have 115 bones; all American eels (*Anguilla rostrata*) have only 107.

529. How do eels reach ponds and lakes which have no connection with rivers?   Eels found in fresh-water rivers, lakes, and ponds are usually females. The males remain near the ocean at the river mouths and normally grow to only about a foot, only rarely reaching 18 inches. The females migrate upstream to feed in fresh water, entering many lakes and ponds from river tributaries. There they have been known to grow to more than 2 feet in length. Biologists have found that eels can survive for several hours out of water and can reach ponds and lakes that appear to have no inlet or outlet by traveling over damp rocks and through underground waterways.

530. Are eels found only in the Atlantic Ocean and streams emptying into it?   Correct. Eels are found in America from northern Canada to Panama and the West Indies, but never on the Pacific west coast. Sea lampreys are often found in salmon streams, but lampreys are not eels.

531. What is the DSL?   The deep scattering layer (DSL) is a widespread layer of living organisms that scatter or reflect sound pulses. During the day, this layer has been reported at depths of 700 to 2,400 feet, but most often between depths of 1,000 and 1,500 feet; at night the layer moves to or near the surface. Existence of the DSL has been reported from almost all deep-ocean areas.

532. What evidence is there that the DSL exists in the Arctic?   Until recently there was a question as to whether the deep scattering layer existed in the Arctic. Well-authenticated reports from Danish and Canadian scientists have now answered the question in the affirmative. Captain H. Worm-Leonhard of the Royal Danish Navy has reported that while surveying Greenland waters he remembers having noted the presence of scattering layers on several occasions, "The first time was in Skovfjorden, in July 1959, where we thought at first that we were grounded on 300-meter water. It was proved by lead that there was no bottom. The second time was in Godthaabsfjorden where, in June 1960, we suddenly got an absolute flat bottom on about 500 meters. Later on we got the normal bottom on 600–680 meters."

P. C. Jollymore of the Atlantic Oceanographic Laboratory, Bedford Institute, Nova Scotia, reports that "Echos from the deep scattering layer stopped the counter during one series of tests, giving the impression of shallow water. This was overcome by increasing the receiver send time after transmit, thereby blanking the return from the deep scattering layer."

The Danish surveyor who participated in the Canadian ice-survey reported that "They had to be careful all the time, lest they get a false echo from the scattering layer which exists all over."

533. What organisms form the DSL?

The mystery of what animals make up the DSL has been largely cleared up by direct observations from submersibles. Siphonophores and fish have been directly identified. Siphonophores are gelatinous coelenterates related to jellyfishes. They possess an air-filled float which acts as a sound scatterer. Other animals which may be significant members of the deep scattering layer are shrimp, euphausiids (shrimplike crustaceans), copepods, lanternfish, and squid.

534. Why does the DSL move upward at night?

The siphonophores expand their floats and move to the surface at night to feed on planktonic animals which are most common in the surface layers. They come closest to the surface on moonless nights. Other organisms also rise to the surface at night to feed on the plankton.

535. Why is it difficult to photograph or net the organisms in the DSL?

Although the organisms are good sound reflectors and may appear on the sonar screen as a "phantom bottom," the individual organisms may be yards apart. Both photographs and net hauls were inconclusive in identifying the organisms. Certain identification has come only through direct observation from submersibles.

536. What did Darwin write about bioluminescence?   In his book *The Voyage of the Beagle,* Charles Darwin wrote: "While sailing south of the Plata one very dark night, the vessel drove before her bow two billows of liquid phosphorus, and in her wake she was followed by a milky train."

537. What is bioluminescence?   Bioluminescence, or living light, in produced by both animals and plants. In contrast to incandescent light, high temperatures are not necessary; oxygen, however, is essential to the light-producing process. The term *phosphorescence* is sometimes used for this phenomenon because it was at first thought to be caused by the element phosphorus in the water. Displays are seen most frequently in warm surface waters. Luminescence is commonly observed as a bluish-green fluorescent flow in water disturbed by the passage of ships or by waves. Occasionally displays are in the form of parallel bars or "wheel spokes" of pulsating light extending from horizon to horizon.

538. What causes luminescence in the sea?   Dinoflagellates are the major cause of surface luminescence. They are single-celled organisms having some of the properties of both plants and animals. They are unique in using the sun's energy for photosynthesis and expending energy in the same form, as light. The light is created by oxygen combining with a substance called *luciferin;* the complicated chemical reaction produces light without heat.

539. How many types of animals are bioluminescent?   About 240 genera, including thousands of species, of marine animals have been identified as producing bioluminescence. In addition to single-celled animals, various jellyfishes, copepods, and euphausiids produce displays. Among vertebrates, luminescence is found only in certain fishes and sharks.

540. Are luminescent organisms found at all depths?
Light-producing organisms have been recorded at all levels down

to 3,750 meters (12,303 feet), and it is probable that they are found even deeper. Beebe estimated that 96 percent of all the creatures brought up by nets were luminescent. In addition to the lower forms of life, perhaps one half to four fifths of deep-water fishes produce light.

541. What are the uses of light organs? There is controversy among biologists concerning the purposes of lights on marine animals. Some animals have well-developed eyes but no light to enable them to see in the dark; others have brilliant light organs but are too blind to see. The property of luminescence is perhaps used as a defense against predators or a means of attracting their prey or a recognition signal for finding members of the opposite sex in the dark.

542. How does bioluminescence aid fishermen? Schools of fish sometimes leave a trail of luminescence behind them. The sardine and anchovy fisheries off California are carried on in the dark of the moon when schools of fish can be located by luminescence.

543. What is the largest shark? Of the 250 known species of sharks, the whale shark is the largest, reaching lengths to 50 feet. A 38-foot specimen caught off Florida was estimated to weigh 26,600 pounds. White sharks have been measured at 30 feet and have been estimated at 40 to 45 feet.

544. How do sharks regulate their buoyancy? Sharks do not have swim bladders, but they have large livers. A 21-foot white shark weighing 7,100 pounds had a liver weighing 1,005 pounds. Scientists of the Seattle Laboratory of the National Marine Fisheries Service have obtained experimental evidence that the liver regulates buoyancy by increasing or decreasing the amounts of diacyl glyceryl ethers, which are significantly lighter than the normal triglycerides.

The sand tiger shark swallows air, thus making use of its stomach as a buoyancy tank.

545. How do sharks find their food? Sharks can pinpoint the sound of a struggling fish or threshing swimmer from distances greater than 200 yards; this is far beyond the visual range.

546. Are all sharks meat eaters?   All sharks are carnivorous except the basking or whale shark, which subsists on plankton. Even this planktonic diet is largely animal.

547. What are maneater sharks?   This name is applied to the great white shark, *Carcharodon carcharias.* The largest specimen ever caught was 36½ feet long. It is most abundant in tropical waters, but has been found almost as far north as Newfoundland.

548. What economic value do sharks have?   Shark livers are rich in vitamins, particularly those of the soupfin, *Galeorhinus zyopterus,* and the dogfish, *Squalus acanthias.* During World War II, when the supply of Norwegian cod-liver oil was cut off, livers of the soupfin shark sold for as much as 14 dollars a pound. The development of synthetic Vitamin A has eliminated the commercial shark fishing business.

549. Are shark repellents effective?   During World War II more than 70 compounds were tested as shark repellents. The most effective were those containing copper acetate, which apparently causes a mucous formation in the nostrils and causes the shark to lose interest in food. Compounds which were effective repellents for some Atlantic species had no effect on many Pacific species.

550. Where do most shark attacks occur?   Most recorded attacks have been in tropical and subtropical areas between the hours of 3:00 and 4:00 in the afternoon; this is the time when there are the most swimmers. Areas which are particularly dangerous are Australia, South Africa, and the Pacific coast of Panama. The Zambezi shark in the South African and Australian waters is considered the most dangerous shark. An almost identical shark, the bull shark off the southeastern U.S. coast, is much less aggressive; aggressiveness in sharks has been correlated with waters of low salinity.

551. How many kinds of sharks are dangerous?   There are approximately 250 species of sharks in the oceans. It is difficult to identify the species involved in attacks on swimmers. At least a dozen species have been positively identified and more than 50 species are considered dangerous.

552. When was the first shark attack in American waters recorded? The first recorded attack was by a hammerhead shark off Long Island in 1815.

553. How can swimmers be protected from sharks? The eastern coasts of South Africa and Australia are two of the most dangerous areas in the world. As many as a dozen shark attacks took place every year in each locality until nets were installed in recent years. The nets have proved very effective; nets are now being installed along beaches in the Gulf of Mexico and elsewhere.

554. How many marine species are harmful to man? More than 3,000 species of marine animals are capable of harming man. There are three categories of dangerous animals—carnivorous, venomous, and poisonous. Carnivorous animals include the shark, barracuda, killer whale, and moray eel. Venomous animals are particularly dangerous to waders. These include the sting ray, stonefish, toadfish, zebra fish, sea urchin, poisonous corals, *Conus* snails, and sea snakes. Poisonous animals are primarily fishes that cause sickness or death when eaten.

555. What is the deadliest animal in the ocean? A 5-inch jellyfish which drifts with the currents off northern Australia has caused more deaths than sharks. The box jellyfish—or sea wasp—has tentacles nearly 25 feet long which contain a cobra-like venom. Swimmers brushed by its tentacles usually die within five minutes.

Another contender for the title of most dangerous animal is the blue-ringed octopus (*Octopus maculosus),* which is found in nearshore waters off Queensland and Sydney, Australia. Although this animal is rarely longer than 10 centimeters (about 4 inches), it carries enough toxin to kill ten men.

556. How did the sea wasp get its name? The scientific name for the sea wasp is *Chironex fleckeri.* The name *sea wasp* was given to this jellyfish because of the very severe and often fatal stings inflicted. Scientists report the venom to be one of the deadliest known; even when diluted by 10,000 parts of water, an injection can cause the death of a laboratory animal in seconds. The venom causes a breakdown of the blood cells in the victim and an asphyxial death.

557. What are the most dangerous animals other than sharks?   The barracuda is feared more than sharks by West Indian divers. Its usual length is only 4 to 6 feet, but it is aggressive, fast, and armed with a combination of long canines and small teeth capable of cutting as cleanly as a knife. Although no authentic record of deliberate attacks on man exists, the killer whale is potentially more dangerous than either sharks or barracudas. This carnivore measures 15 to 20 feet and hunts in packs. It attacks seals, walruses, porpoises, and even baleen whales.

558. Why is the moray eel dangerous?   The moray eel, which is as long as 10 feet, lurks in holes in coral reefs and may inflict severe lacerations on a diver who pokes his hand into its hiding place, or it may grasp the diver in its bulldoglike grip until he drowns.

559. Is the giant clam really dangerous?   *Tridacna gigas* grows to a length of 4 feet and weighs several hundred pounds. Divers have been reported trapped and drowned by this mollusc. The danger, however, has been exaggerated. They live on shoal reefs in clear tropical waters and can easily be seen because of their large size, colorful mantles, and habit of squirting water at low tide. A diver trapped by the clam can free himself by inserting a knife between the valves and cutting the two muscles that hold the halves together.

560. How dangerous is the octopus?   The only thing fierce about the octopus is its appearance. The animal is timid and will usually hide in a hole at the approach of a swimmer. The greatest danger to humans is not the tentacles, but poisoning from a bite.

561. What is the holding power of an octopus?   Strength varies with the size, maturity, and physical development of the octopus. The largest species may weigh as much as 50 pounds and have tentacles up to 10 feet long. Each sucking disc can apply 4 ounces of holding pressure. Multiplying this value by the 2,000 or more sucking discs that a 50-pound octopus may have, provides a total holding pressure capability of 500 pounds or more.

562. How might a diver best free himself if he becomes entangled with an octopus?   Any diver who carries a knife with him, as he should, when diving, has the proper tool for securing his release from an octopus. The procedure does not, however, begin with cutting off the tentacles. Specialists in diving safety say that destruction of the nerve center is the quickest and surest way to obtain release. The brain of the octopus is located between and above the eyes, and should be the target for the knife. Until this nerve center is destroyed, suckers and tentacles will continue to operate effectively, regardless of other physical damage.

563. Are sea snakes deadly?　All sea snakes are venomous and many are deadly; they are closely related to cobras. Their habitat is shallow coastal waters and coral reefs; they are most abundant in tropical waters. Sea snakes can be distinguished from terrestrial snakes by their paddle-shaped tails. Their bodies, unlike those of eels, are covered by scales. Sea snakes have a mild disposition and will seldom attack humans without provocation. Most fatalities result from fishermen catching the snakes in nets and contacting them unintentionally.

564. How much electricity does an electric eel generate?　Although the electric eel (which isn't a true eel) is the best known generator of electricity, there are at least 500 kinds of fishes that generate appreciable amounts of electricity. The electrical discharge serves to stun prey and repel attackers. The average discharge is more than 350 volts, but discharges as high as 650 volts have been measured. Current is low, usually a fraction of an ampere; however, brief discharges of 500 volts at 2 amperes have been measured, producing 1,000 watts. Although direct current is produced, it may be discharged as frequently as 300 times a second.

Severity of the shock depends on the size and state of health of the eel. Voltage increases until the eel reaches a total length of about 3 feet; after that, only amperage increases. Electric eels in South American waters have been known to grow to a length of almost 10 feet.

565. Are giant devil-rays dangerous?　The giant devil-ray, or *manta,* has a span up to 20 feet and may weigh more than a ton. Despite its size and sinister appearance, it is not armed with a stinger and is not aggressive. The smaller sting-ray may inflict a serious wound on a wader who steps on it.

566. Are any snails dangerous?   The cone shell (*Conus*) is the only snail that inflicts injury on humans. The attractive 4-inch shell has sharp spines with venom glands at the base of each spine. Shell collectors picking up specimens of *Conus* may be temporarily paralyzed by the venom; there are several reports of collectors who have died from cone-shell stings.

567. What is the Portuguese man-of-war?   *Physalia* is a jellyfish having venom almost as virulent as that of the cobra. Although the sting can disable a man for days, there are no records of fatal stings. The Portuguese man-of-war received its name from its iridescent float, which resembles a sail.

568. Are any fishes dangerous to eat?   Not all fishes are edible. Some have organs that are always poisonous to man; others sometimes become toxic because of certain elements in their diet. There are 300 tropical species of fishes that cause fish poisoning; one type of poisoning is known as *ciguatera*. A particular species may cause ciguatera when caught on one side of an island, but not if caught on the other side. These tropical fish are associated with reefs and do not usually venture far from the home reef; for this reason, the people living on one island may eat a certain species of fish, while those on a nearby island would not. No one knows what causes the fish to become poisonous, but most investigators agree that it is something in the diet. There is no method to determine before a fish is consumed whether or not it will cause ciguatera. Some common species of fish known to cause ciguatera are: surgeon fish, jacks, porgies, snappers, goatfish, moray eels, wrasses, and barracudas.

569. What is scombrid poisoning?   Scombrid fish, commonly known as tuna or mackerel, have been known to cause scombrid poisoning, usually because of inadequate preservation. The flesh of scombrid fish contains bacteria which, if the fish is not preserved soon after capture, begins to produce a histamine-like compound. This compound, if ingested by humans, causes a severe allergic reaction and may even lead to death.

570. What is fugu?   In Japan a national dish called *fugu* is highly prized. It is prepared from the puffer fish, and the gonads of the puf-

fer are highly poisonous. For this reason, fugu is only served in res-
taurants licensed by the government.

571. Is it safe to eat sharks? Consumption of sharks and rays
has been known to cause illness or death; this was probably because
the victim ate a portion of the liver, which contains a very high concentration of vitamin A that the human body cannot tolerate.

572. Are barracuda edible fish? In general, marine biologists
and fishery authorities do not recommend eating barracuda. The
chances are that one will become quite ill if a barracuda longer than
14 inches is eaten, and there is no assurance that smaller ones may
not be toxic. Three species of barracuda inhabit waters off the southeast coast of the United States: the sennet, the gauguanche, and the
greater barracuda. The greater barracuda has a wide oceanic range
and is known to be poisonous in some regions. The toxicity of the
other two species is questionable. The poisonous condition of barracuda flesh is the result of their feeding on smaller fishes that eat toxic
materials. Toxic materials are stored by lower organisms or fish and
passed on and stored in the flesh of the barracuda. In more tropical
waters where coral and other related marine life become more abundant, the chances of catching barracuda that have fed on poisonous
marine life increases substantially.

573. What mammals live in the sea? Members of at least four
orders have adapted to life in the sea. The Pinnipedia have flippers,
or *pinnas*. They include seals, walruses, and sea lions, totaling as
many as 25 million individuals. The Sirenia include the sea cows,
manatees, and dugongs. Sea otters are the major representative of the
Carnivora. The Cetacea include whales, porpoises, and dolphins. Although not usually considered a marine mammal, polar bears are
well adapted to the water.

**574. How long has it been known that whales are not
fish?** Aristotle noted that not only whales but also porpoises and
dolphins are mammals rather than fish, as was generally believed before his time.

**575. Why do whales, dolphins, and porpoises have horizontal
tail fins?** Marine mammals must come to the surface to breathe.

The horizontal tail is useful in bringing its nostrils above water; the nostrils are located on top of the head rather than on the snout.

576. Are whales found in all oceans of the world?   Most of the larger species continually migrate from ocean to ocean with the changing seasons and have been observed in all oceans and even in fresh water. There are two distinct species populations: the Southern Hemisphere group and the Northern Hemisphere group. Both groups breed in tropical coastal waters during the winter, then go to the Arctic or Antarctic regions for summer feeding. The fin, sei, and humpback whales migrate seasonally in this manner.

577. How large do whales grow?   At birth blue whales may be 26 feet long. Thirteen years later they reach maturity and a length up to 100 feet; weight may be 150 tons. Sperm whales average 14 feet at birth and mature more slowly.

A blue whale's heart weighs half a ton and generates 10 horsepower to pump its 8 tons of blood.

578. How long do whales live?   The best evidence of age of whales is the successive layers (laminae) of teeth and ear plugs. There is not yet agreement among scientists as to whether more than one lamina is formed in a year, but it is assumed that a fin whale can live about 80 years.

579. What do whales eat?   The toothed whales catch fish and squid. The cachalot, or sperm whale, belongs to this group. Each tooth weighs a half pound or more.

Whalebone, or baleen, whales consume small planktonic animals. The blue whale has a throat so small that it can swallow only small animals; the daily consumption is probably several tons.

580. How fast can a whale swim?   The blue whale, which may be almost 100 feet long, can reach speeds of 20 knots for a few minutes. It can maintain a cruising speed of 15 knots for hours. Dolphins, which are small whales, can reach peak speeds of 25 knots.

581. How deep do whales dive?   Whales that feed on plankton have no reason to dive below 300 feet because their food is concentrated in the upper layer of the ocean. The toothed whales dive much deeper in search of squid and fish. Whales sometimes become entangled in cables on the bottom, furnishing exact information on the depth of their dives. The deepest recorded entanglement was 3,720 feet.

582. Why don't whales get the bends? Toothed whales may dive very rapidly to a half-mile depth, from a pressure of one atmosphere to nearly 80 atmospheres. Because the whale has only the air in its lungs—rather than the unlimited supply available to a diver— very little nitrogen is dissolved in its blood and tissues.

583. Do whales ever sleep? Because whales are mammals they must rise to the surface periodically to breath. They must also maintain the proper volume of air in their lungs to maintain buoyancy at the surface. Because of this they probably do not sleep deeply as humans do. When whales are immobilized by drugs they either sink and drown or stop breathing and suffocate.

584. Where do sulfur-bottom whales get their color? The sulfur-bottom whales of the Antarctic are colored on the underside by yellow diatoms, single-celled plants.

585. What whale is most sought by whalers? The fin whale, or common rorqual, has become the mainstay of the whaling industry because the blue whale has been hunted almost to extinction. As whaling becomes more efficient the fin whale is being seriously reduced in numbers. International agreements between whale-hunting nations attempt to limit the kill of certain species of whales. Through this international effort it is hoped that the whale stocks will be preserved for future generations.

586. What are right whales? These were considered by whalers to be the right ones to catch, not only because they produced the most oil and whalebone, but also because they were buoyant and could be more easily towed to the ship or to port. Three types of right whales are found off the shores of the United States: the bowhead or Arctic right whale, the Atlantic right whale, and the Pacific right whale.

By the beginning of the twentieth century right whales had become so scarce they faced extinction. By international agreement in 1937 they were protected from hunting except by aborigines in need of food; since then they have begun to increase slowly.

587. What is whalebone?   Whalebone whales have no teeth, but have baleen plates hanging from their upper jaw to strain out small animals for food. The plates are tough and pliable; in the past they were used for whalebone corsets.

588. What is ambergris?   This substance, valued from 2 to 9 dollars an ounce, is formed only in the intestine of the sperm whale. How it is formed is not known with any certainty. The material, resembling a black rock, is sometimes ejected by the whale and is found by beachcombers. More often it is found by the whalers. Its main use is as a fixative in the manufacture of expensive perfumes. Chunks of ambergris vary from less than an ounce to almost a thousand pounds.

589. Are dolphins mammals or fish?   Both; the dolphin, *Delphinus,* is a small whale and *Coryphaena,* the dolphin fish, is a bright-colored fish that lives in tropical and subtropical waters and feeds on flying fish.

590. What   is   the   difference   between   dolphins   and porpoises?   Although very similar, dolphins and porpoises are frequently placed in different families by zoologists, chiefly on the basis of the form of their teeth and the presence of small bony protuberances on the forward edge of the dorsal fin. Dolphins usually have long, beaklike snouts; the snout of porpoises is short and blunt. The names are often used interchangeably; the large gray mammal that is trained to jump and perform tricks is frequently called a porpoise but is actually a bottlenose dolphin.

591. Is it true that porpoises attack sharks?   Yes, porpoises are the natural enemies of sharks. A school of porpoises will surround a shark and butt it with their hard noses. The continued butting ruptures the shark's internal organs and kills it.

592. How do porpoises get fresh water? Being mammals, porpoises have the same need for fresh water that humans have. It is assumed that their fresh water comes from the fish and squid they eat.

593. How fast can a porpoise swim? Most porpoises can swim 17 to 23 miles per hour for short periods, although, to an observer aboard a ship, they may appear to be traveling much faster. There are records of porpoises being observed at 40 to 43 miles per hour, but they were swimming before a ship, utilizing the bow wave for extra speed. Much research has been done to discover just how the porpoise is able to accomplish its high swimming speed. Either it is a much more powerful swimmer than expected, or it modifies its shape and, therefore, reduces hydrodynamic drag. The question is yet unsolved.

594. Is the porpoise the fastest swimming animal? Although the porpoise is a very fast swimmer, it is not the fastest sea animal. Marlin, bonito, and albacore have been reported to swim at speeds of 40 to 50 miles per hour. The sailfish and swordfish have attained speeds of 60 miles per hour.

595. How sensitive is a porpoise's sonar? Tests conducted by the U.S. Navy proved that porpoises can distinguish between aluminum and copper discs and could usually distinguish size differences of one quarter inch in steel balls the size of a golf ball.

596. What is the manatee? This is the fabled mermaid of old sailors' stories. Its forelimbs are absent, with the body ending in a broad, flattened tail. It lives in shallow coastal waters and eats only aquatic plants. Manatees inhabit the warm rivers of Florida and estuaries of the Indian and Pacific Oceans on both sides of the equator.

597. How does the sea otter differ from the land otter? They are very similar in appearance, except that the sea otter is larger and better adapted for marine living. It has webbed hind feet, and its fur is said to be the world's best. At one time sea otters were very common, but they approached extinction because of the high demand and value of their skin. In 1911 an agreement known as the "Fur Seal

Treaty" between the United States, Russia, Great Britain, and Japan protected fur seals and sea otters from further hunting or trapping. This treaty is still in effect and violation carries a very severe penalty.

598. What does the sea otter eat?   The sea otter feeds on mussels and other shellfish and has a very interesting feeding behavior pattern. It floats on its back and pounds shellfish against a rock balanced on its chest; the pounding fractures the shell and enables the sea otter to feed on the soft parts.

599. How are seals classified?   Seals are divided into two groups, those with external ears and those without. Eared seals include the sea lion, fur seal, and walrus. Earless seals comprise the harbor seal and the elephant seal.

600. Where is the fur seal found?   It is found all over the world. Great herds spend the summer months in the Bering Sea and Pribilof Islands. The fur seal is small, weighing only a few hundred pounds, and its fur is very valuable. It has been protected by treaty since 1911; only 3-year-old bachelors are taken each year for furs.

601. What is the habitat of the walrus?   It inhabits the Arctic and adjacent seas. Walrus have a thick layer of skin and fat for protection from the cold water and two large tusks for digging shellfish and crustaceans from the bottom.

602. What are the earless seals?   Earless seals include the harbor seal and the elephant seal. The harbor seal is smaller than the eared seals; it grows to 5 feet in length and weighs about 150 pounds. It feeds on fish and crustaceans. Usually harbor seals stay close to land, mainly around harbors and river mouths. They range from the coasts of the United States to the Arctic Circle.

603. What is the largest seal?   The elephant seal is the largest of all seals. The male is considerably larger than the female and may have a body length of 16 feet and a weight up to 5,000 pounds. The elephant seal was so named because of the large snout of the male.

Distribution ranges from California to Alaska and they are also numerous in the North Atlantic.

Largest of the eared seals is the northern sea lion, weighing as much as 1,700 pounds. The California sea lion weighs only about 600 pounds; the female has a fine sense of balance and is trained for circus work.

604. What air-breathing animals are adapted to diving? Aquatic animals include mammals, birds, reptiles, and amphibians. Among mammals that dive can be listed many well-known fresh-water types ranging from the small muskrat and water shrew to the otter and beaver and the unusual Australian platypus. Some of the large diving mammals, the hippopotamus and the manatee are found in both fresh and salt water; man too dives in both fresh and salt water. Diving mammals found in ocean waters include porpoises, whales, seals, walruses, and polar bears. Listed among the diving birds are diving ducks, penguins, puffins, cormorants, and guillemots. There are a number of diving reptiles including salamanders and sea snakes; probably the most impressive in this category are marine turtles and crocodilians.

605. What pressures can diving mammals withstand? The following table lists the depth that diving mammals are known to have dived to, based on extensive investigations of marine biologists and physiologists:

Otters	to		60 feet
Bottlenose dolphin. . . .	to		65 feet
Fur seals.	to	about	250 feet
Steller sea lions	to	about	500 feet
Most seals	to	about	850 feet
Weddell seals	to	about	1,150 feet
Sperm whales	to	about	4,000 feet

606. Which of the diving animals is known to have the longest breath-holding capability? To this question most of us would answer, "The whale, of course." According to exhaustive studies conducted by physiologists, we find that whales stay underwater from 60 to 120 minutes before returning for another breath of air. The marine turtles are able to dive or stay underwater much longer; many in captivity are known to stay underwater for more than 3 hours before returning to the surface for air. One authority states that some species of aquatic turtles can stay submerged for periods up to several days; he has recorded four species of turtles that are able to extract measurable quantities of oxygen from the water while submerged. This may account in part for the unusually long breath-holding capability of certain turtles.

607. How does the breath-holding capability of man compare with other mammals?

Most men	1 minute
Polar bears	1.5 minutes
Pearl divers (humans)	2.5 minutes
Sea otter	5 minutes
Platypus	10 minutes
Muskrat	12 minutes
Hippopotamus	15 minutes
Sea cow	16 minutes
Beaver	20 minutes
Porpoise	15 minutes
Seals	15 to 28 minutes
Greenland whale	60 minutes
Sperm whale	90 minutes
Bottlenose whale	120 minutes

608. Do diving mammals have greater lung capacity than terrestrial animals? Generally speaking, yes, diving mammals have slightly greater lung volume, in addition to greater blood volume and oxygen storage capacity than nondiving species. There is, however, one notable exception, the whale. If lung volume is measured in terms of its ratio to body weight, then its lung capacity is only about one-half that of terrestrial mammals. It is important to note that the whale's method of breathing is different from most diving mammals. Whales use a much larger part of the total lung capacity in normal respiration than does man. Whales breathe deeply with each breath, whereas man normally does not. Each breath for a whale might be equivalent to hyperventilating breathing for man.

609. Are there any common physiological characteristics that enable diving animals to stay underwater for prolonged periods? George Washington University biologists have recently completed a study of past work done on this subject. Investigators report that there is usually little difference in lung volume between diving and terrestrial animals, but a significant physiological difference is that most diving animals have a greater potential to store oxygen in their blood. Another difference is that diving animals seem to be less sensitive to carbon dioxide buildup and oxygen lack; i.e., they can tolerate a greater carbon dioxide excess and utilize more completely the oxygen supply in the lungs. Another important physical characteristic common to most diving animals is the pronounced slowing of the heart; this occurs at the beginning of a dive and continues until the animal surfaces. Naturally there are other changes in the circulatory system that occur as an adjustment to the slowed heart beat.

610. How are coral reefs formed? Living corals build on the skeletons of their predecessors; the living corals are usually found only at the upper and outer edges of the reef. Coral may constitute less than half of the reef material.

611. What forms other than coral contribute to reefs? The main reef builders are algae; without them coral could not build reefs. The framework of the coral-algae reef is filled in with shells of clams, oysters, snails, and echinoderms. Even worms add to the reef by building calcareous tubes.

612. How do light and temperature affect reef building? The single-celled alga which supplies the coral with calcium carbonate and consumes the coral's waste products requires light for growth. The optimum depth for the algae is about 15 feet but they can grow at depths of 150 feet if the water is clear. The extreme temperature range of coral polyps is 67° to 90° F; they thrive best in water between 77° and 86° F. The salinity range is 27 to 40 parts per thousand; because of this they do not develop near the mouths of large rivers.

613. Why are coral reefs formed only in warm-water areas? Calcium carbonate is much more soluble in cold water than in warm. This may explain why only isolated corals are found in cold water while large reefs develop in tropical waters.

614. What do corals eat?   Coral polyps use the stinging cells on their tentacles to paralyze small planktonic animals. They then bring the food to their mouths with the tentacles.

615. How fast does a coral reef grow?   Rate of growth depends on the type of coral and environmental conditions. A typical growth rate for a reef or coral boulder might be an inch or two a year.

616. Where are coral reefs found?   Most are near the equator in the Pacific and Indian Oceans. One of the most extensive, the Great Barrier Reef of Australia, is more than 1,200 miles long and up to 95 miles wide.

617. Why does coral occur off the Florida coast and not off California?   Coral is found only off the east coasts of the world's continents, where the prevailing winds and the earth's rotation push warm tropical waters toward the poles. Coral-reef builders reject building sites along the deeper western continental shores where cold upwelling currents, which flow toward the equator, preclude their growth.

618. How are coral atolls formed?   Charles Darwin proposed the idea more than a hundred years ago that the sea bottom has been subsiding slowly over the ages. As volcanic islands went down, corals built upward. When the islands disappeared entirely, nothing was left at the surface except a circular coral atoll with a lagoon in the center.

619. How has Darwin's theory been proved?   During Darwin's lifetime there was only circumstantial evidence to support his theory. Chunks of dead coral were brought up from far below the limiting depth of coral growth. On some coral islands, native statues were standing in water, indicating that subsidence was still going on. In addition, the geographic distribution of atolls indicated a broad regional subsidence; the central part of the atoll area apparently subsided too rapidly for atolls to build up.

Then, in 1952, a hole was drilled on Eniwetok atoll through more than 3,000 feet of coral, into the basaltic rock of an extinct volcano. Darwin's theory was proved.

620. What starfish destroys coral reefs?   The Crown of Thorns starfish, *Acanthaster planci,* was a minor predator of coral reef invertebrates until recent years. Beginning in 1966 it multiplied rapidly; by 1969 it had destroyed hundreds of square miles of coral reef in the Great Barrier Reef and elsewhere in the Pacific. In two years more than 90 percent of the 24 miles of coral on the northwest side of Guam was destroyed.

621. What is the Crown of Thorns?   This starfish has many arms extending from a central disc; it can grow as large as 18 inches in diameter. It eats live coral by extending its digestive sac over coral formations and digesting the living coral. The remaining coral skeleton can eventually break up, destroying the reef.

622. What caused the Crown of Thorns population explosion?   Man-induced pollutants and the results of dredging and blasting have upset the ecological balance and reduced the natural enemies of this starfish. One of the natural enemies, the triton snail, has been reduced in numbers by shell collectors.

623. What other animals destroy reefs?   Worms and snails bore into coral, killing the polyps. At least one variety of sponge can dissolve coral. Many fishes feed on the coral polyps by biting off chunks of the reef and digesting the polyps. The parrotfish scrapes the reef with its beak. The bumpfish (*Bolbometron*), which can weigh more than 60 pounds, breaks off chunks of coral with its head and then chews them, consuming several thousand pounds in a year.

624. What are shipworms?   They are one of the most destructive groups of marine borers; other destructive groups are crustacean borers and rock borers. Two shipworm genera, *Teredo* and *Bankia,* grow to more than 5 feet, with a diameter less than an inch. A considerable amount of research is being conducted to determine how they may be controlled. It is known that they are sensitive, in varying degrees, to salinity, temperature, food supply, pH, and the amount of dissolved hydrogen sulfide in the water.

625. How much damage can marine borers cause? Between 1917 and 1921, borers caused failure of almost every underwater wooden structure in the northern arm of San Francisco Bay. Damage was estimated at $25 million.

In a 10-year period ending in the late 1930s, the crustacean borer *Limnoria* caused damage of $5 million in Boston Harbor.

626. What catastrophic failure was caused by borers? In 1946 the Brielle Bridge over the Manasquan River in New Jersey collapsed as a result of the activities of marine borers in the central bridge supports.

627. Are there any kinds of woods that are naturally resistant to marine borer attack?   In the absence of borers the durability of wood underwater is high. But when exposed to borers, wooden structures suffer severe damage, especially in lower temperate and tropical waters. In U.S. waters, marine borer damage is estimated at 200 to 250 million dollars annually. Pressure treatment of wood with whole creosote is the most commonly used method of borer protection, but where borer infestation is high, especially in tropical waters, even creosoted timbers require frequent replacement. Some woods such as Greenheart and Angelique have been commercially marketed on the basis of being borer resistant. In order to check the reputed resistance of certain woods, particularly native tropical species, the Panama Canal Company and the Naval Research Laboratory (NRL) have completed a 7½-year test of 115 wood species at two ocean sites and a brackish-lake site in the Canal Zone. The samples were evaluated for resistance to the three principal borer classes, teredo, pholad, and limnoria. The test results indicate (1) that limnoria work rapidly in only a few woods, notably U.S. conifers. Pholad populations develop slowly but eventually damage most woods. (2) The brackish-water teredo species *Psiloteredo healdi* was generally more destructive than all the 28 ocean borer species combined. (3) Wood density does not seem to be a controlling factor in borer resistance except possibly with pholads. However, dense woods were generally more resistant to all borers than lighter species. (4) Wood with high silica content was usually effective in resisting ocean teredos but silica content seemed to have little influence on resistance to pholads and limnoria. (5) Many woods were selectively resistant to certain borer species. Two woods with excellent structural properties were found to be resistant to all ocean borers except pholads. These were Laurel and Cedro Espino. In northern waters where pholads are not a problem these woods should be excellent marine materials. (6) One wood was found to be highly resistant to all classes of borers at each of the three environmental sites. It is Cocobolo, but its small size and the irregular shape of the tree make it unsuitable for marine construction timbers. Its high borer resistance is not due to its silica content, which is very low, but the oily constituent of the wood, which was found toxic to the skin of some individuals, is thought to be the protective component. Scientists expect to identify this substance and are hopeful of developing a similar synthetic material for treating other timbers for marine use.

628. What is fouling? The mass of living and nonliving bodies and particles attached to or lying on the surface of a submerged object. The term is usually applied only to living bodies attaching to man-made objects.

629. What kinds of animals foul ships? One of the most common and tenacious is the barnacle. This small animal attaches itself to anything man-made in the water and stays until it is scraped off. The larger mussels are common foulers. Bryozoans form plantlike fans.

Various species of hydroids, calcareous worms, and sea squirts are free-swimming in their larval stage; they attach themselves to the hulls of ships in port and remain in place unless removed by friction of the moving ship.

630. What makes barnacles adhere to rocks, ships, and other marine structures so tightly? Scientists at the University of Akron have recently discovered that barnacles secrete a special cement, sometimes called *barnacle glue,* which hardens in 10 or 15 minutes when put in water. According to laboratory strength tests, a layer of this barnacle glue 3/10,000 of an inch thick has a shear strength of over 7,000 pounds per square inch (about twice the strength of the best epoxy glues). In these investigations, the scientists found a series of paired glands located at the inner base of the shell, containing two fluids, one milky white, the other a pale brown. The milky fluid is thought to be the shell-building material and the brownish substance the cement. It appears that these glands operate in pairs and new ones are continually grown to replace the old, which dry up once the new shell is added and secured.

631. Is fouling greater in warm water? Fouling organisms thrive best in warmer waters. However, in colder waters, mussels can grow to tremendous proportions in a very short time. The problems, though different, are not always less in cold water.

632. How fast does fouling occur? After six to eight months in the water, a ship may accumulate 2 to 3 inches of growth, weighing as much as 100 tons, particularly if the ship has spent some time in tropical ports.

633. Does fouling occur at all depths? Fouling organisms are found in the shallow waters along shore and also along the bottom in the deepest parts of the oceans. However, they are not always found continuously from the surface to the sea floor.

634. How does fouling affect a ship's speed? When a ship's hull is heavily laden with barnacles, top speed may be reduced by 2 knots. The friction caused by fouling organisms can so reduce the speed developed at a given engine power that, in order to maintain shipping schedules, fuel consumption must be increased by 50 percent.

635. How does marine life affect the conduct of engineering operations? The insulation of electrical cables may be penetrated by organisms and mooring cables may be severed. Moving parts of a deep-sea device may be immobilized by incrustations of fouling organisms. One of the major limiting factors of oceanographic instruments designed for extended underwater operation is fouling, which can cause the instruments to malfunction; current meters become inaccurate and acoustic receivers become less sensitive.

636. What materials are not affected by fouling? Materials showing little or no biological deterioration include various types of rubber, nylon, Teflon, and glass. Fouling organisms do cause deterioration of natural fiber, cable insulations, and untreated wood.

637. How were wooden ships protected from marine organisms? Copper sheathing was used for protection against boring and fouling organisms because of its toxic properties. By 1783 all English vessels were copper-sheathed, and by the early 1800s the French and Spanish had followed suit. Copper sheathing has now been replaced by coatings and paints, many of which contain copper. Because the toxic material must dissolve fast enough to prevent attachment, these coatings must be renewed periodically.

638. How did oceanographers assist in the development of antifouling coatings? During World War II, oceanographers of the Woods Hole Oceanographic Institution worked with the U.S. Navy to learn how marine paint actually works and which compounds are most effective at the least cost. Their research saved millions of dollars by cutting the cost of paints, lengthening the stay out of dry dock, and saving fuel. The Navy attributed a 10-percent fuel bill reduction to the improved antifouling paints.

639. What are the newest antifouling materials? Toxic additives are being used with rubberoid compounds. Organo-tin additives have shown a greater toxicity and longer effective life than the older copper-base paints. Plastic coatings containing ions of mercury can poison organisms within 1 millimeter of a ship's hull and prevent attachment of larval organisms.

640. Can fouling organisms live in fresh water? Few fouling organisms are found in fresh water. If ships remain in fresh water for a sufficient period of time, fouling organisms will be killed; contrary to popular belief, the fouling organisms do not drop off the hull. When the ship returns to salt water, the layer of dead organisms provides an excellent base for the attachment of new organisms.

641. Can fouling organisms survive a trip through the Panama Canal? Professor R. J. Menzies of Florida State University towed a mesh bag filled with fouling organisms through the Panama Canal to determine whether they could survive. Barnacles, shipworms, crabs, and snails were all able to survive the trip.

642. What special fouling problems do supertankers have? Drydocks for supertankers are few and far between. Therefore, slow-growing organisms which present no problems to ships scheduled for regular drydock maintenance are a potential hazard. Although most plants and animals are repelled or killed by the poisonous action of copper compounds on hulls, the brown alga *Ectocarpus siliculosus* can tolerate 0.1 mg per liter of copper and continue to grow.

643. Are there bacteria in the sea? They are as abundant in the sea as on land; the ocean contains about 1,500 different kinds of bacteria. There may be as many as a million in a cubic centimeter of seawater.

644. Where are bacteria most abundant? They are most abundant at the bottom, where they serve as scavengers. In turn, they serve as food for protozoans, worms, and sponges. Bacteria are found in all parts of the ocean from surface to bottom. Free-floating species are most common in the near-surface zone of photosynthesis.

645. Have any insects been identified as marine insects? To date, only one marine insect is known to exist; it is a creature commonly known as the water strider. The scientific name is *Halobates*. The water strider was first found when Kotzebue circumnavigated the globe aboard the *Rurik* in 1815–16; at that time, Eschschole described 3 species of the insect. An extensive treatise on the *Halobates* was done in 1888 after the *Challenger* Expedition added several species to those already known. By 1969 a total of 39 species had been identified; 6 of them in the past decade.

646. Is the water strider a true insect? Yes; the water strider (*Halobates*) belongs to the order Hemiptera. They have 6 legs and have lost their wings but are extremely well adapted to the marine environment. The middle and back legs are elongated and well adapted for moving rapidly over the water. The middle pair are designed for rowing or pushing and the rear legs serve as rudders. The body of the water strider is covered with fine hairs which repel water just as feathers do for a duck. This prevents them from getting wet if submerged. *Halobates* is found in all oceans in tropical and warm temperate waters.

647. What does the water strider feed on? Water striders are predatory by nature and feed on the body fluids of victims by puncturing them with their piercing and sucking mouth parts. A variety of marine organisms living on or near the sea surface serve as the diet of these insects; frequently jellyfish such as the Portuguese Man of War (*Physalia*) fall prey to the water strider.

648. What is known of the life cycle of Halobates? Some species of the water strider live near the coast; others spend their entire life far from shore. These insects are gregarious, collecting in large numbers near floating objects. Eggs are laid on driftwood, bird feathers, or any floating object. They develop from egg to adult in about 2 months. Some scientists suspect that in the nymphal stages it swims underwater; many details of the life cycle are as yet unknown. Efforts to study *Halobates* in an aquarium have been unsuccessful; in captivity they strike the aquarium wall because of their rapid movements and die as a result. This being the case, it is uncanny how they can survive in the open ocean, withstanding waves of gale force strength; this is another problem for study. They are limited geographically to ocean areas where surface water temperatures are 21° C (69.8° F) or higher.

649. How far from the shore are land insects found? As recently as 1968 it was the general consensus of marine biologists that land insects could be found some distance at sea only under rare conditions. However, results obtained from cruises of the R/V *Crawford* and R/V *Gosnold* in the summer of 1968 changed that concept. Common land insects were collected during the *Crawford* cruise at a distance of 150 miles offshore. These insects included moths, butterflies, stinkbugs, ladybugs and other beetles, dragonflies, many kinds of small flies and aphids. Another unexpected find of the *Gosnold* cruise was the discovery that terrestrial insects made up over half of the food of the lantern fish, *Gonichthys coccoi,* one of the most abundant midwater species of the western North Atlantic. These light-producing fish make regular nightly vertical migrations to the surface waters to feed.

650. Are penguins only found in Antarctica? There are 17 species of penguins, all of which live in the Southern Hemisphere. Seven species of this aquatic bird are found in Antarctica. The two most southern species are the Emperor (*Aptenodytes forsteri*) penguin and the Adélie (*Pygoscelis adeliae*) penguin. The Emperor is the larger bird; an average adult stands about 3 feet tall and weighs about 60 pounds. Emperors nest on sea ice that is attached to the land. In autumn, a single egg is laid. For protection from the winter cold, the egg (or young chick) is held between an adult's feet and a roll of stomach fat. When the chicks are able to feed themselves, the colonies—called rookeries—break up and the Emperor penguins move north in search of food.

Adélie penguins are about 18 inches high and weigh about 14 pounds when full grown. They lay their eggs, usually two, in the spring after building a nest of pebbles; by summer's end, the chicks are able to take care of themselves.

Other species of penguins are found in the South Temperate Zone as far north as Patagonia, South Africa, Australia, and New Zealand.

Biologists have tracked these flightless birds returning to their rookeries from great distances, but how the penguins navigate is still being studied.

651. What other penguins inhabit Antarctic waters? In addition to the Adélie and Emperor penguins which seldom migrate beyond 60° S, there are five other species that live in Antarctic waters. These are:

The Gentoo (*Pygoscelis papua*) which is sometimes called the Johnny; adults reach a length of 2½ feet and weigh 8 to 14 lbs.; they are found as far as 65° S on Palmer Penninsula.

The Ringed (*Pygoscelis antarctica*) adults are 2½ feet in length and 7 to 12 pounds in weight.

The Rockhopper (*Eudyptes crestatus*) adult length is about 2 feet. It has the unusual trait of jumping like a kangaroo when in a hurry;

also it jumps feet first into the sea from ledges, not diving as the Adélies do.

The Macaroni (*Eudyptes chrysolophus*) adult length is about 2½ feet and resembles the Rockhopper.

The King (*Aptenodytes patagonica*) adults are 3 feet in length and weigh about 40 pounds. King penguins are indifferent to human beings; as a result they have been almost exterminated by man for their oil.

652. What do penguins eat?   The penguin depends on a seafood diet which is mainly fish, squid, and crustaceans (shrimplike creatures). The King penguin prefers mainly squid and feeds closer to shore. The Adélie seems to prefer crustaceans and the Emperor is somewhat less selective. When on the pack ice which is farther from the Antarctic Continent, the Emperor must depend on very small fish and planktonic crustaceans for sustenance.

XI. ECOLOGY

653. What is ecology?   The Council on Environmental Quality defines ecology as the science of the intricate web of relationships between living organisms and their living and nonliving surroundings.

654. What is meant by the term ecosystems?   Ecosystems might best be defined by illustrations; forests, lakes, and estuaries are examples of ecosystems. These are interdependent living and nonliving parts that function as identifiable natural units.

655. What are biomes?   Biomes are larger ecosystems or combinations of ecosystems which occur in similar climates and share similar character and arrangement of vegetation. Examples are: arctic tundra, prairie grasslands, and the Sargasso Sea.

656. What is the biosphere?   The biosphere consists of the earth, its surrounding envelope of life-giving water and air and all living things.

657. Are there seasons in the ocean?   In winter the surface waters are thoroughly mixed by storms. Nutrients are brought to the surface. With the coming of spring, diatoms make use of the nutrients to increase rapidly. They, in turn, serve as pasture for planktonic animals. In summer, the surface layers are warm and nutrients are no longer brought to the surface; plants are consumed faster than they can grow. In autumn, plankton bloom because storms stir up subsurface nutrients which have formed during the summer from decaying plants and animals.

658. How do fish react to seasonal temperature changes?
Temperature in the ocean does not change as rapidly as temperature in fresh water, nor is the range as wide. Therefore, marine fishes have less tolerance to changes. Some fishes migrate southward in the winter; others move into deeper waters where seasonal temperature changes are minimal. Some become inactive or bury themselves in the mud.

659. Are there temperature barriers in the oceans? The Cape Cod area is one example. Water temperature south of the Cape is about 10° F warmer than that north of the Cape. Many species are found only north of the Cape or only south of it. Others may be found in both areas, but breed in only one area.

660. Is there more life in cold or warm water? There are fewer kinds of animals in the Arctic than in warmer areas, but they are larger and each species is represented by more individuals. Both oxygen and carbon dioxide are more soluble in cold water than in warm water. There are, in general, more nutrients available in the Arctic areas to provide food for plants which serve as the primary food for animals.

661. How well can estuarine fishes adapt to temperature change? Fishes in estuaries are more tolerant to changes in temperature than those in the open ocean, but less tolerant than those in fresh water. Many species cannot tolerate temperatures greater than 90° F, which is approached on a hot summer day in the lower latitudes. Even a slight rise in temperature caused by industrial discharge or cooling water from a nuclear power plant could be fatal.

662. Do environmental conditions have any effect on the growth rates of lobsters? Results of recent experiments in both the United States and Canada have proved that increased temperatures promote faster growth. For example, the four larvae stages are normally completed in 30 days at 55° F; at 62° F these stages are completed in 20 days and at 70° F, in 10 days. Higher temperatures also produced better survival rates through the larvae stages. Faster growth through juvenile and sub-adult stages has been observed at warmer temperatures, but no conclusions have been reached as to the optimum temperature for the best growth rate.

663. What effect does temperature have on lobster fisheries? The lobster catch on the Maine coast dropped from 24½ million pounds in 1957 to less than 20 million in 1966. During these years the average water temperature decreased from 49° F to 45°.

664. How does salinity form barriers in estuaries? Some fresh-water animals, such as the gar, cannot move from river water into the brackish water of an estuary. Others, such as the fresh-water snail, can move freely into estuaries. Oysters, crabs, and shrimp are adapted to wide salinity variations. Mussels have a narrow range and starfish cannot tolerate less than 15 parts per thousand salinity.

665. Why is life less abundant at the bottom of the ocean? One reason is the scarcity of food. Even if food were available, however, the low amount of dissolved oxygen would be a limiting factor.

666. Why is oxygen important in seawater? Dissolved oxygen is important in the metabolism of organisms. It is also an essential element in the formation and solution of lime and in the decomposition of organic matter.

667. How is oxygen added to the sea? In the surface layer, oxygen is absorbed from the air. From the surface to the depth of light penetration, oxygen is produced by photosynthesis of plants. Mixing of surface water and convection move oxygen-rich water to lower depths.

668. What causes hydrogen sulfide to concentrate at the bottom of the Black Sea? The large rivers emptying into the Black Sea form a low-salinity layer that is lighter than the underlying water. This upper layer contains oxygen, but because there is no vertical circulation, the bottom water does not. In the absence of oxygen, sulfate-reducing bacteria produce hydrogen sulfide from the sulfate in seawater.

669. What are the effects of hydrogen sulfide? Hydrogen sulfide is not only lethal to most forms of life but is also highly corrosive to many materials. The blackening of white lead paint is a well-known phenomenon in badly polluted estuaries.

670. What pH range can fish tolerate? In general, fish can live in the pH range of 5 to 9. However, changes in acidity may make several common pollutants more toxic. Fish mortalities may be expected below a pH value of 5.0, but some fish can adapt to a pH as low as 3.7.

671. Why do animals make daily vertical migrations? Dr. George L. Clarke of the Woods Hole Oceanographic Institution has suggested that animals migrate vertically in order to seek a certain light intensity. Some invertebrates rise at night to feed on plants and are followed by fish and squids which feed on them. Diurnal migrations to the surface may be from depths as great as 1,200 feet.

672. Do fish migrate through the Panama Canal? No; the fresh water in Gatún Lake is a barrier to most organisms. If, however, a sea-level canal is built across the isthmus, fish and other organisms could move freely between the Atlantic and Pacific.

673. What ecological changes would be caused by a sea-level Panama Canal? Ecologists are concerned about the possible changes in fish populations that could be brought about by environmental changes or introduction of new predators. Water on the Pacific side is cooler than the Atlantic; the Pacific also has higher tides. Flow of colder water from the Pacific to the Atlantic could alter the environment.

There are a number of closely related species on opposite sides of the isthmus which have been separated by the land barrier for 80 million years. If these species are brought together they might interbreed and form a hybrid population or they might remain separate. Some species might be eliminated; others might become more abundant.

674. What predator might enter the Atlantic through a sea-level Panama Canal? Sea snakes, related to cobras and every bit as deadly, inhabit the Pacific side of Central America. A sea-level canal might enable them to migrate to the Atlantic side. If sea snakes entered the Caribbean in large numbers they might upset the ecological balance; they might also upset the tourists.

675. Have animals migrated through the Suez Canal? Migrations have been mostly from the Red Sea to the Mediterranean; few animals have migrated the other direction. In the hundred years since the canal was opened, 126 animals and 4 plants are known to have migrated successfully. About 25 species of Red Sea fishes have migrated to the Mediterranean; two of these now constitute Israel's primary fishery. Shrimps and crabs are other economically important animals which have migrated to the Mediterranean.

676. What trace elements are essential to growth of marine plants? Copper, zinc, molybdenum, and cobalt are all essential for growth of plants. There is also evidence that growth of plants is limited by deficiencies of iron.

677. What elements are concentrated by organisms? Copper is concentrated by many organisms, including oysters and crustaceans. Some elements cannot be detected in seawater, but are concentrated in organisms. Vanadium has been found in the body fluids of tunicates and holothurians, nickel in molluscs, and cobalt in lobsters and mussels. Iodine and bromine are concentrated in the brown seaweeds.

In recent years there has been considerable concern about concentration of radioactive elements and mercury in organisms used for food.

678. How do organisms affect the composition of seawater? Nearly all the dissolved salts are used for nourishment, but in very different proportions. In areas of intense biological activity, substances such as phosphates and nitrogen compounds may be reduced below the normal level. Decomposition of plants and animals also alters the chemical composition locally.

679. How is carbon dioxide produced? The carbon dioxide which is needed for plant photosynthesis is produced by animal respiration; it is also dissolved directly from the atmosphere.

680. What are the nutrient elements? Potassium, calcium, and magnesium are always present in amounts which exceed the needs of marine plants. Nitrogen, phosphorus, and silicon are not always present in overabundance; plant growth under conditions favorable to photosynthesis may deplete the supply.

681. How are nutrients replenished in the ocean? If agriculture is to be highly productive, plant nutrients must be replenished through artificial fertilization. In the ocean, nutrients are replenished by natural processes such as microbial action and inflow of streams and rivers carrying agricultural fertilizers and sewage. Plants and animals sink after death and decompose to add nutrients to the water.

682. How are chemical nutrients changed into food?   Through photosynthesis, the phytoplankton changes silicates, nitrates and phosphates into primary food which is used by the zooplankton. This is the first step in the food chain.

683. What is photosynthesis?   Photosynthesis is the process by which simple sugars and starches are produced from carbon dioxide and water by living plant cells with the aid of chlorophyll and in the presence of light. During this process, the plant takes up carbon dioxide from the water and releases oxygen into the water. Photosynthesis by aquatic plants, including algae, is an important source of oxygen in water.

684. What is the food chain?   The food chain involves all the marine plants and animals and even the chemical constituents of seawater. Minute plants float near the surface, utilizing the sun's energy to synthesize organic matter from dissolved nutrients. Tiny zooplankton eat the phytoplankton and are in turn eaten by larger animals. For example, copepods, which are crustaceans about the size of a grain of rice, eat tiny floating plants. The copepods are eaten by herring, which may be eaten by larger fish or by man. There are three different food chains: the *predator* chain, the *parasite* chain, and the *saprophytic* chain.

685. What is the predator chain?   The initial level is the plants, followed by the herbivores, followed by various levels of carnivores. Each level of similar size is called a *trophic* level.

686. What is the parasitic chain?   In this chain each successive level is smaller in size because they exist at the expense of larger organisms.

687. What is the saprophytic chain?   Saprophytes live on dead organisms. Plants and animals which are not eaten during their lifetime die and become food for scavengers or they decay, aided by bacteria. The nutrients released by decay return to the plants, completing the food cycle.

688. How much plant life is required to produce a pound of fish?   About 100 pounds of phytoplankton are needed to produce 10 pounds of zooplankton, which in turn produce 1 pound of herring. Larger and more desirable fish higher up the food chain, such as salmon, would gain only ounces from the original 100 pounds of plant material.

689. How can the food chain be shortened?   One method would be to harvest plankton. However, fish are much more efficient plankton collectors than are men. By fishing for the plankton eaters, such as herring and anchovy, the yield would be much higher than it now is. Blue whales live almost entirely on *krill,* a shrimplike animal abundant in the polar seas. With the decrease in whale catch, the Russians began harvesting krill and processing it for human food.

690. What are metabolites?   These organic particles are secreted or excreted from plants and animals or are the products of decay of dead organisms. They cling to air bubbles and may be seen as a thin brownish film on the sea surface. As they sink they are eaten by the zooplankton. They may also be absorbed by plants.

691. What factors restrict the mobility of fish?   Mobility of many fishes is limited by barriers of temperature and salinity, as well as availability of food. Some kinds of fishes are adapted only for life on the bottom; others are so sensitive to pressure changes that they are limited to a narrow vertical zone.

692. How do fish at great depths withstand pressure?   Even the pressure of 8 tons per square inch at the greatest depths is not a limiting factor to fish because the pressure is equalized throughout the tissues. However, sudden changes in pressure can be lethal. Deep-sea fish brought to the surface rapidly may have their swim bladder expanded out through their mouth.

693. Can fish change their color?   Some flatfishes can expand and contract their color cells to blend with the bottom on which they live. It is known that they make a visual observation of the bottom; blinded fish cannot change color.

694. Why isn't there a fish population explosion? Codfish produce 5 million eggs at a time. If all of them survived the Atlantic would be filled solidly with cod in 6 years.

A female mackerel produces half a million eggs, but after 62 days only 20 specimens will have survived. By the 85th day after spawning only 2 will be alive.

695. What ecological factors affect the abundance of fish? Physical conditions include temperature, currents, bottom conditions, and light. Chemical conditions include salinity, oxygen content, and presence of nutrients.

696. How does light affect fish populations? The annual cycle of light intensity affects the daily cycle of fish behavior, maturation of gonads, and metabolic changes. The amount of light which penetrates suspended particles and reaches the lower layers of water affects orientation of fish in relation to food, depth, and predators. In addition, light is essential for the growth of plants which serve as food.

697. How does temperature affect fish populations? Water temperatures affect the spawning and migration of fish. Fish species have certain temperature ranges which are characteristic for adult species, for their eggs and for their young. Alterations in temperature can be harmful to adults or young.

698. How have temperature changes affected the sardine fisheries? Before World War II the sardine fishery off California was the largest fishery in the United States. By 1951 the catch had dropped from 500,000 tons to 3,000 tons. During the years of declining catch, the water temperature had decreased. In 1957 and 1958 coastal water temperatures increased 2 to 4 degrees and there was an immediate increase in the sardine catch.

699. How do water temperatures affect tuna fishing? The northern and southern range of various species of tuna is limited by temperature, and their seasonal migrations are related to the seasonal changes in water temperature. Ninety percent of bluefin tuna are caught in 62° to 70° F water, beginning in late May off the coast of southern Baja California. The seasonal appearance of the skipjack tuna in the vicinity of the Hawaiian Islands is associated with the time and space shifts in the California Current extension. Fishermen make use of sea-surface temperature charts, transmitted by radio, in scouting for tuna.

Scientists of the National Marine Fisheries Service Center at La Jolla, California, have examined the log records of over 2,000 sets of purse seines on bluefish tuna. They found 36 percent greater success in water between 59° and 65° F, as compared with water between 70° and 76° F.

700. How does water temperature affect the fisheries of Peru?   The anchovies which have made Peru one of the world's leading fishing nations are quite sensitive to water temperature. They thrive in the cool upwelled water of the Peru Current, between 14° and 18° C (57.2° and 64.4° F). The anchovies move vertically and inshore-offshore with the shift of waters of these temperatures. In years when warm currents replace cold currents, the catch is severely reduced.

701. What effect do winds have on fishing?   If fishermen have information on the seasonal and short-term changes in prevailing wind direction they can decide whether it is worth while to go to a particular area. For example, cod fishing in the Barents Sea is good when the winds are from the southwest; these winds drive more warm water into the fishing area and the cod follow it. When the winds are from the southeast, fishing is poor; the wind impedes the flow of warm water and allows colder water to enter the Barents Sea.

702. How can research on ecology help fishermen?   An understanding of the ecology is the basis for maintaining populations that will produce maximum yield year after year. Research on the habits and reactions to changing sea conditions can provide the basis for developing means to catch little-used populations cheaply.

703. What are the desert areas of the ocean?   The regions of really blue sea are almost devoid of life. In the tropics the sea is relatively stable, and the chemical nutrients tend to sink below the level of light penetration. Therefore, photosynthesis proceeds at a slow rate even though light penetrates deeply into the clear water. The Sargasso Sea is a typical desert area.

704. Can marine organisms be transplanted from one ocean to another? Successful transplantations have been made of pink salmon from the North Pacific seas of the USSR to the North Atlantic seas of the USSR. Atlantic oyster culture in Nantucket and Martha's Vineyard and importing Japanese seed oysters to the Pacific Northwest are other examples. But for every successful transplantation there are numerous failures.

705. How can transplanting organisms endanger the ecology? Transplanted species can seriously disturb the ecological balance to the detriment of harvestable species already present. Predatory Japanese snails introduced into the Black Sea in 1949 virtually eliminated mussel populations and apparently caused a sharp decline in flounder fisheries. A great number of inadvertent introductions into the Atlantic and Pacific Oceans could result from opening the proposed sea-level canal across Central America.

706. What are sea lampreys? The adult sea lamprey is an eel-like parasite that averages about 15 inches in length. It preys on fish, attaching itself by a circular suction-cup-shaped mouth which has sharp rasping teeth that bore in the flesh and allow the lamprey to feed on the victim's blood. The sea lamprey has existed and preyed on fish in Lake Ontario for hundreds of years and apparently gained access to the other Great Lakes when the Welland Canal was deepened between 1913 and 1918. By 1956 it had virtually destroyed the lake trout fishing industry in Lakes Michigan and Huron. In 1958, scientists discovered a chemical that was highly effective against young lamprey which hatch and live in tributary streams before becoming parasites and migrating into the lakes. These measures introduced in Lakes Superior and Michigan have resulted in an 80 to 90 percent reduction in lamprey population, according to a 1969 report.

XII. FOOD FROM THE SEA

707. What areas of the world ocean provide the greatest volume of the commercial fish catch? According to fisheries experts of the National Marine Fisheries Service, 80 percent of the volume (and 90 percent of the value) of the world catch of marine fish is taken over the various continental shelves where the water is less than 100 fathoms (600 feet) deep.

708. What countries catch the most fish? In 1969, Japan passed Peru to become the world's leading fishing country. Peru's fisheries are based primarily on anchovies, which have decreased in recent years. Japanese fishermen caught 9.2 million metric tons during 1969. The Soviet Union was in third place with 7.3 million tons. Mainland China is believed to be in fourth place, although statistics are not available to the Food and Agriculture Organization of the United Nations.

709. How does the fish catch of the United States compare with other countries? In 1956 the United States ranked second in the world in fish catch; by 1966 the United States had dropped to sixth place. In 1969 the United States moved up to fifth place because a shortage of herring in European waters decreased Norway's catch.

710. What is the value of the U.S. catch? The value of the U.S. catch, excluding shellfish, has been about half a billion dollars a year since 1969. The average price for all species received by fishermen in 1969 was 12 cents per pound.

711. What is the U.S. consumption of fish? Although the United States is not generally considered a fish-consuming country, the per capita consumption of fish is nearly double the world's average. Much of the catch is eaten by livestock, poultry, and pets. The United States consumes about 12 percent of the world's catch, making it the world's largest importer. In 1969 about 60 percent of the total U.S. supply of fish products came from imports.

712. Why has the U.S. fish catch not kept up with the demand? Part of the reason is that the United States is able to buy any fish it needs. Other nations take more fish from traditional U.S. fishing grounds than do U.S. fishermen. However, except for a few fisheries, like tuna and shrimp, U.S. fishermen are not in a competitive position. The American fleet is technically outmoded. High construction costs make American fishing boats expensive.

713. Why are fish an important food in Japan? Although fish make up less than 3 percent of the calorie consumption in Japan, they furnish 74 percent of the animal protein. On a worldwide basis, fish account for 12 percent of the animal protein.

714. Are whales still caught by U.S. fishermen? In early 1971, Secretary of Commerce Maurice H. Stans directed the National Marine Fisheries Service—the agency responsible for administering whaling regulations—to terminate the United States licensing of commercial hunting of whales listed as endangered species. The list contains 8 species, including the fin, the sei, and sperm whales, which are the only species sought by U.S. commercial whalers in recent years. Other species on the endangered list were bowhead, blue, humpback, right, and gray whales. The whales were placed on the endangered species list by the Department of Interior in December 1970. This action marked the final chapter in an industry which originated along the New England coast in the early colonial days, and reached its highest peak during the first half of the nineteenth century. The domestic whaling industry had been on the decline since the beginning of the twentieth century, but for the first time in almost three hundred years, no whaling ships will be operating from what is now the United States coast.

715. What was the U.S. whale catch in recent years? According to the National Marine Fisheries Service, the 1969 U.S. whale catch was reported to be 108 animals. The breakdown by type was: 31 fin, 10 sei, and 67 sperm whales. This latter figure included 34 undersized sperm whales taken under a special scientific permit for study by scientists of the Bureau of Commercial Fisheries (now National Marine Fisheries Service) marine mammal biological laboratory.

716. How large is the U.S. squid catch?   Although squid have never attained the popularity in the United States that they have in the Orient, there is a surprisingly vigorous commercial fishery centered on the California coast. The National Marine Fisheries Service reports an annual catch of more than 20 million pounds, valued at over half a million dollars, for the last several years.

Earliest statistics available indicate a catch of 30,000 pounds in 1895. The record of 38 million pounds was reached in 1946.

717. Why does India use so little of the abundant fish crop in the Indian Ocean?   One of the major reasons is the lack of refrigeration which is so necessary in hot climates. Canning of fish or use of fish protein concentrate could provide an adequate diet to the undernourished population around the Indian Ocean.

718. Why is food from the sea becoming increasingly important?   Since the end of World War II, there have been such significant advances in world health that the populations of emerging nations have increased more rapidly than their economic development. As a result, nearly two thirds of the world population suffers from protein malnutrition. Population expansion has, at the same time, reduced the amount of land available for grazing animals which could supply the vitally needed animal protein. The need for vast quantities of low-cost animal proteins has forced man to turn to the sea.

719. What is fish protein concentrate?   It is a substance which is tasteless and almost odorless but contains all the amino acids essential to humans in proper proportions to maintain health. In concentrated form, fully dehydrated and defatted, it can be shipped and stored for long periods without refrigeration. Roughly refined fish meal has been used as feed for chickens, pigs, and cattle, but it was not until February 1, 1967, that the U.S. Federal Food and Drug Administration approved the use of whole fish protein concentrate for human consumption. Ten grams will provide adequate animal protein to meet the daily requirements of one child. Efforts are under way to reduce the cost of this daily requirement to less than one cent.

720. How long has fish flour been in use?   There is nothing new about the idea of fish protein concentrates for the direct human diet. They have been used since antiquity wherever fish have been available; they were used in the Far East and the Roman Empire more than 2,000 years ago. A fish flour for use in biscuits was produced in Norway in 1876. Very high-quality fish flour was made in Germany during World War II and used to replace egg white in baking. In South Africa, fish protein concentrate was used to enhance the protein quality of cereal products in 1937 and then abandoned for lack of a market. Production was resumed in 1951 and by 1958 over a thousand tons had been used in enriching brown bread.

721. How important is fish meal as an animal food? Chicken has become one of the cheapest meat products on the market because of the widespread use of fish meal for feed. If fish meal were not available in large quantities at low cost, the hog-raising industry would be at a severe disadvantage in competing with other meat products.

722. Why is fish protein concentrate made from whole fish? Whole fish are used primarily for economy. There are, however, other benefits. Proteins and calcium in the bones and phosphorus in the skin add to the nutritional value. In August 1970 the U.S. Food and Drug Administration approved the use of herring and menhaden in producing FPC. Previously only hake and "hake-like" fishes were approved.

723. How much of the fish catch is processed into meal? More than one third of the world fish catch is converted to meal and used for animal food. The total catch of Peru, which exceeds all other countries except Japan, is turned into fish meal and oil.

724. How is fish oil used? Fish oil is especially valuable as a base for heat-resistant paints. Other uses include the manufacture of soap, candles, margarine, cooking fats, varnishes, waterproofing compounds, and linoleum.

725. Can fish be increased by fertilizing the sea? Adding agricultural fertilizer to the sea would increase the plankton on which animals feed. In one cost study, it was estimated that doubling the nutrients of the North Sea would cost at least ten times the value of the additional fish which could be harvested. Although fertilizing the open ocean is not economically attractive, bays and estuaries appear somewhat more promising.

726. Can hatcheries increase fish populations? Since the greatest fish mortality occurs between hatching and the end of the larval stage, it seems reasonable to assume that protecting large quantities of desirable species beyond the period of greatest danger would help to increase stocks. Experiments are now being conducted to determine whether stocks will increase.

727. Can fish be transplanted to new locations? Among species which have been successfully transplanted from the east coast to the west coast of North America are the striped bass and shad. The North American Chinook salmon has been transplanted to New Zealand. Only a small percentage of attempted transplants have been successfully established.

Stocks of young fish moved from areas of overcrowding to nearby areas of abundant food have shown dramatic increase in growth. However, international cooperation would be necessary before such transport can be done on a large scale. No nation would finance such moves when all nations could harvest the fish.

728. Can air bubbles be used to fence in fish? It has been demonstrated in Japan, Canada, Norway, and Scotland that fish can be herded by curtains of air bubbles as effectively as by nets, provided that currents are less than 3 knots. Faster currents cause turbulence that can destroy the curtain of bubbles.

729. What types of marine biota are cultivated in coastal areas? Japan probably has the longest history of cultivating marine areas. They cultivate edible marine seaweeds, oysters, clams, and other shellfish, prawns, and several types of marine fin fish. Other countries have proceeded to develop their marine aquaculture programs with these same types. Incidentally, the common carp will live and grow in brackish water with nearly one third of the salinity of ocean water, so this species of fish is used for fresh-water inland ponds as well as coastal brackish ponds like those on the Baltic Sea.

730. What are some of the problems in fish farming? One of the main problems in developing a successful commercial fish culture project is insuring that the water resource will be of the right quality. Usually this means absence of harmful polluting substances and, equally important, making sure that the amount of dissolved oxygen in the water is adequate. A deficiency of dissolved oxygen is a common problem because organic matter often uses it up in the decaying (oxidization) process. The quantity of oxygen needed varies according to the activities of the fish and according to species. Other problems can be pond leakage, unwanted vegetation growth, and run-off with much silt in suspension. For marine animals, regular renewal of water of proper salinity levels must be accommodated in ponds.

731. When was aquaculture started? Aquaculture is not a modern innovation at all, though the word *aquaculture* is a term thought to be coined by Dr. A. F. Spilhaus and used by him in an address given in the mid-1960s. One form of aquaculture, fish farming, is known to have been practiced in China more than 3,000 years ago (1100 B.C.). Culture of the common carp (*Cyprinus carpio L.*) was the subject of a Chinese book written about 460 B.C. It indicates that the carp was the first of the world fishes to be domesticated from Chinese rivers for "cultivation." During the Tang Dynasty (A.D. 618–907) four other local river fishes were added to Chinese fish culture; these have since then been the basis of fish farming in China. Fall stocking of ponds and moats was practiced in several European countries during the Middle Ages as a means of providing food for the winter and might be considered an early form of fish culture.

732. How much fish and other seafood is cultivated? Results of a survey completed recently by the United Nations Food and Agriculture Organization (FAO) indicate that approximately 4,000,000 metric tons of fish and shellfish are produced throughout the world annually. This figure is based on averages of national production reported in recent years.

733. Which countries are the leaders in aquaculture production? According to the UN's Food & Agriculture Organization (FAO), countries of Asia and the Far East are by far the greatest cultivators of fish and shellfish animals for food. The annual world production of cultivated fin fish in the late 1960s totals 3,-000,000 metric tons. Mainland China was first in a list of 36 nations with a production of 1,190,000 metric tons. Japan was second with 487,000 metric tons; India was next with 480,000 tons and was followed by the USSR with 190,000 tons.

734. Where are the major fishing grounds? Most fishing grounds are in coastal areas in water depths less than 1,200 feet. Shallow areas contain more food because the plants which furnish food and oxygen can grow only at depths where sunlight penetrates. Additionally, rivers carry nutrients from the land to coastal waters.

Areas around the British Isles, Iceland, the Faeroe Islands, and off the coasts of Norway, Russia, and Newfoundland are among the most heavily fished in the world. Other well-established fisheries are found around Japan, off South Africa and Malaya. In recent years fishing has increased off the coasts of South America.

735. Where is the richest fishing area in the world? Waters off the coast of Peru are probably more productive than any area of similar size in the world. The anchovetas which are caught for fish meal and oil make Peru one of the leading fishing nations of the world. Although fishermen catch more than $100 million worth of fish every year, the coastal birds eat many times more. They, in turn, produce guano worth about $20 million a year.

736. Why are most fish caught in the Northern Hemisphere? Although there is much more water in the Southern Hemisphere, an estimated 98 percent of all edible fish taken from the oceans are caught in the Northern Hemisphere. One reason is that the large population centers are in the Northern Hemisphere; another reason is that there are more productive fishing grounds. Because there is more land in the Northern Hemisphere, there are also more coastal and continental shelf areas which are feeding grounds for fish. Currents and upwelling along coasts stir up nutrients which sustain fish populations.

737. How long ago did commercial fishing begin in American waters? As long ago as 1550 a thousand vessels were fishing for cod with long lines on the Grand Banks. Portuguese fishermen still fish for cod in this area in spring and early summer, preserving the catch with salt.

738. What makes the Grand Banks a good fishing ground? The cold Labrador Current meets the Gulf Stream in this area, stirring up chemical nutrients which fertilize the phytoplankton, the minute drifting plants which are the food of small animals. Herring eat the small animals and are in turn eaten by cod. In recent years Soviet trawlers have been catching herring in this area. Although this may reduce the population of cod, it may also increase the population of haddock. The herring eat great quantities of haddock eggs.

739. What are banks? This is the term used by fishermen for shoal areas surrounded by deeper water. The discovery of new seamounts and their charting by marine geologists has led to the discovery of good fishing spots. Fishermen use bathymetric charts and echo sounders to locate the spots of greatest fish concentration.

740. What is the importance of estuaries to fisheries? Coastal and estuarine waters and marshlands furnish the nutrients and spawning grounds for two thirds of the world's entire fisheries harvest. Seven of the 10 species most valuable to American commercial fisheries spend all or important periods of their lives in estuaries, and at least 80 other commercially important species are dependent upon estuarine areas.

741. What are the most abundant fish? Herring are the most abundant fish in the world. For centuries they have been caught in the greatest numbers. They are especially abundant in the North Atlantic and are an important item in European diets. The herring family includes pilchards, anchovies, sardines, and menhaden.

742. How many kinds of fishes are commonly used as food? Of the thousands of species of fishes in the oceans, less than a dozen make up more than half of the world's catch. The most important food fish are the herring, cod, mackerel, salmon, flatfish, and rosefish.

743. What fish did the early colonists use for fertilizer? The Indians taught the American colonists to place a menhaden in each hill of corn they planted. Menhaden are members of the herring family and have been used in great quantities for fish meal, oil, and fertilizer. In 1889 more than 173 million pounds of menhaden were landed in New England. During the early part of the twentieth century they failed to come north of Cape Cod, possibly because of low water temperatures. Since 1950 they have returned to New England waters and are again an important commercial fish.

744. What are pelagic fishes? Pelagic fishes are found in the upper layers of the sea. Some live offshore part of the year, moving inshore to spawn. Such are the sea herring, mackerel, and menhaden. Others, including the tuna, are true inhabitants of the open ocean, coming close to land in their extensive movements about the sea.

745. What are groundfish? Groundfish live on or near the sea bottom in the cool waters over the continental shelf. Examples are haddock, flounder, cod, and whiting. One of the world's largest groundfisheries is in the Northwestern Atlantic, including such well-known fishing grounds as Georges Bank, Browns Bank, the Nova Scotian banks, and Grand Banks of Newfoundland. Most of the catch is filleted and sold fresh or frozen; very little is salted or canned; and some, such as the whiting (silver hake), go to reduction plants to be made into fish meal and concentrates.

746. What are anadromous fishes? Anadromous fishes spend part of their life in the open ocean. In spring they move inshore, entering the bays and mouths of tidal rivers. They spawn upstream in fresh water before they return to the sea. Examples are salmon, shad, and alewives.

747. What is a fish ladder and what is its purpose? A fish ladder (or fishway) is a water-filled lock, channel, or series of connected pools by which fish may swim past or around an obstruction, such as a dam. Its purpose is to provide for successful upstream or downstream passage of fish past the barrier. These are most commonly used by salmon in their upstream spawning runs. However, other anadromous fishes might also benefit from fishways; these include ale-wives (buckies), sea-run trout, shad, and sturgeon.

748. How are migrating salmon kept out of power plant intakes? Since 1965 the National Marine Fisheries Service has been developing a device that will divert immature salmon and other anadromous fishes from destruction by power plant turbines. To date, devices such as curtains of rising bubbles, electrical apparatus, lights, louvers, sound, and water jets have been tried, each with limited success. The latest equipment design, which consists of a horizontal traveling screen hung vertically over the face of the intake, appears to be successful in diverting almost 100 percent of the fish. It is expected that general use of this equipment will increase anadromous fish populations.

749. What is the world harvest of food from the sea? The total annual harvest of food from the sea has risen from less than 20 million metric tons in 1948 to more than 64 million tons in 1969. Production of seafood has been increasing faster than the growth of population, but world demand for fish is expected to reach 100 million tons by 1985.

750. How much can the fish catch be expanded? If fisheries continue to restrict their catch to species now being utilized and continue to exploit the same areas, using the same fishing gear, it is doubtful whether production can be increased more than three or four times the present catch. Beyond this point fishing might become uneconomical unless new areas are investigated and additional species are utilized.

751. Is there danger of overfishing? Overfishing is already a problem for some species. Stocks have been depleted in heavily fished areas such as the continental shelves of Europe, particularly the North Sea. Cessation of fishing during two World Wars proved that a decrease in fishing could result in an increase in the number of large specimens. When the catch of a species reaches the point where reproductive capacity is unable to compensate for the losses sustained, the species is headed for extinction. However, before this

point is reached, operation of fisheries becomes uneconomical, and
fishing of many species to extinction is thus prevented.

752. What species are endangered?   The U.S. National Marine Fisheries Service has listed the following species as being seriously depleted: Pacific sardine, Atlantic salmon, Atlantic sturgeon, blue whale, fin whale, Atlantic shad, sperm whale, humpback whale, oyster, and sea otter. Depletion of these species is not caused entirely by overfishing; disease, predators, and water pollution all take their toll.

753. What percentage of the world's food comes from the sea?   At present only 1 or 2 percent of the world's food supply comes from the oceans. Fish provide about 3 percent of man's direct protein consumption, but because fish meal is fed to land animals, fish are the basis of about 10 percent of all animal protein food production.

754. What are the relative meat production ratios of fish and land animals?   There are many different figures cited on the relative number of pounds of fish or land animals produced per acre. Much depends on the land quality and water quality available. However, one significant point made by most authorities is that fish (and other aquaculture) ponds can be developed on poor land or land otherwise unusable for agriculture. Dr. C. F. Hickling, a British fishery authority, gives a figure of 300 pounds liveweight per acre per year as an average production figure for young cattle on "good permanent pasture in England." By comparison, he gives a fish production figure of 2,500 pounds per acre per year for ponds located on poor tropical soils such as those in Malacca.

755. What species are unexploited?   Bluefin tuna and skipjack tuna fishing began about 1962 in the Atlantic. Exploratory fishing indicates considerable potential for tropical tuna species. The catch of jack mackeral off the Pacific coast of the United States is about 45,-000 tons a year; research indicates that greatly increased catches could be sustained. About a million tons per year of anchovies and large quantities of hake could be caught off the coast of California and converted into fish meal. Catching these fish would probably help to rebuild the stock of sardines which compete for food. There are also vast quantities of anchovies off South Africa.

756. What areas are unexploited? Areas off Argentina and southern Brazil are largely untouched, although the fish population is known to be large. The western side of the Arabian Sea is another area which could support large fishery operations.

757. What is fisheries biology? It is the study of the relationship of fish and their environment, with particular emphasis on the ocean conditions that bring about economically harvestable fish concentrations. Fisheries biologists are concerned with the ecology and population dynamics of the fish populations, knowledge that is necessary to maintain maximum sustainable harvests.

758. How does oceanographic knowledge assist in increasing the harvest of the sea? Oceanographic studies provide leads to highly productive new fishing areas and promising underutilized resources in old areas. Fisheries oceanographers provide fishermen with information to improve tactical scouting and catching operations. Forecasts of seasonal and long-term abundance of fish populations are also provided to fishermen. Oceanographic studies also provide the basis for management of heavily exploited fisheries.

759. How many fish can an acre of ocean produce? Fishermen off the coast of Peru are catching more than 400 pounds of anchovies per acre every year. Productive areas of the continental shelf on the east coast of the United States produce only about 20 pounds per acre. The North American fish are carnivorous species used as human food; the anchovies are plant-eating species used for fish meal.

760. Are zooplankton a potential source of food? Baleen whales consume zooplankton, such as the Antarctic krill, in great quantities. With the decline of whaling, commercial exploitation of krill is a distinct possibility.

761. How many fishing boats are there? An estimated 1 million vessels of various sizes and about 4 million men are engaged in marine fisheries.

762. What is the difference between pelagic and demersal fisheries? Pelagic fisheries include species in the upper layers of open water. Examples are herring and mackerel. Demersal species live on or near the bottom. Haddock, sole, and cod are examples.

763. How are tuna caught? Until recently tuna were caught on hooks by fishermen who had to physically heave the fish on deck. Now most tuna are caught by purse seining. When the fish are spotted, motorboats leave the tuna vessel, towing nets which may be half a mile long and 210 feet deep. When the tuna are surrounded, the net is closed like a draw-string purse and the resulting bag is hauled aboard the tuna vessel by winches.

764. What are trawls? They are nets designed to skim along the bottom, catching 3,000 to 5,000 pounds of fish in one tow. Common types are the beam trawl, held open by a beam along the upper edge, and the otter trawl, held open by two boards which are spread apart by the flow of water. Otter trawls and midwater trawls may also be used at intermediate depths.

765. What is electrofishing? In a field of short electric pulses, fish swim toward the anode where they are stunned and float to the surface. Fish herded by a purse seine can be electrified and pumped into a ship's hold.

766. How are lights used to lure fish? Scientists of the National Marine Fisheries Service are using lights to lure fish into captivity. In experiments at Panama City, Florida, underwater structures are used to attract fish during the day. Although the fish normally tend to leave at night, scientists have found that artificial lights can be used to hold the fish at these structures. By slowly moving the lights, the fish can be herded into areas where they can readily be captured by purse seine or other conventional fishing gear. During one trial, more than 10,000 pounds of Spanish sardines were successfully herded and captured by this method.

Japanese fishermen use subsurface lamps leading to the entrance of nets, switching off the lights one by one until the school of fish is led into the net.

767. Can fish be lured by sounds? Fish are attracted by many sounds, including sounds of engines and screws of fishing boats. In many parts of the world, fish are gathered by making tapping, rasping, or other sounds. Fish are also driven into nets by beating the sea surface with bamboo poles.

768. How could satellites aid fishermen? About three eighths of the tuna fisherman's time at sea is spent looking for tuna. If a satellite system could locate the schools, cutting search time in half, the catch could probably be increased by 25 percent per day.

769. What is the normal yield of fish obtained using standard filleting procedures? Fishery processing technologists say that the usual conventional filleting techniques produce edible table meat equal to about 25 to 30 percent of total bodyweight of the fish. Yields using the Japanese fish processing equipment range from 37 to 60 percent of the body weight.

770. Has any commercial equipment or machinery been developed to automatically debone fish? Since the early 1950s, Japan has been using machines to produce boneless and skinless fish. Fishery scientists from Seattle were first introduced to the equipment during a 1968 visit; a similar machine has been obtained by the U.S. National Marine Fisheries Service Laboratory at Gloucester, Massachusetts, and is undergoing tests for use in the United States fishery industry.

771. How does the fish deboning machine work? First the fish must be headed and gutted. It is then fed into the machine and is carried between a belt and perforated drum. Pressure applied by the belt on the fish forces the flesh through the holes in the drum after which the skin and bones are carried to the waste chute. To insure removal of all bony material the fish flesh is processed further through a meat strainer. The product is known as *minced fish* and is free of both skin and bones.

772. What are shellfish? To the shellfish belong such molluscs as the scallop, clam, and oyster, which are enclosed in a hinged double shell; and crustaceans, as the lobster, crab, and shrimp, which are covered by a horny armor that is actually an outer skeleton.

773. What is the most valuable seafood resource in the United States?
The shrimp industry is by far the most valuable fishery in the United States, having replaced salmon and tuna. In 1967 the shrimp fishery exceeded $100 million and accounted for about 25 percent of the value of all domestic fishery resources. It created more wealth than salmon and tuna combined.

774. How long does it take shrimp to reach a marketable size?
The National Marine Fisheries Service, Galveston, Texas, laboratory reported in 1969 on an experiment in which white shrimp were grown in a pond located at the laboratory. The growth rate was determined by weight. When placed in the pond the shrimp were about half an inch long; more than 43,000 were required to weigh a pound. At the end of one week 3,027 weighed one pound; at the end of three weeks 329 weighed one pound and at the end of 5 weeks, 79 weighed a pound. At the end of the 5-week period the shrimp were 4½ inches in length.

775. How much shrimp does the United States import?
Although the United States catches more shrimp than any other country, since 1961 this country has imported more shrimp than it produces. Imports come from 67 countries and account for almost 25 percent of the edible seafood imports. Mexico is the main exporting country. India and other Asian countries are exporting shrimp in increasing quantities.

776. How fast is the shrimp fleet growing?
Because the demand for shrimp continues to exceed the supply, construction of shrimp trawlers has increased rapidly. More than half of all fishing vessels built in the United States in 1968 were shrimp trawlers, an addition of 350 vessels.

777. Where are the major U.S. shrimp grounds?
Coastal areas from North Carolina to Texas account for more than 80 percent of U.S. shrimp production. Landings of shrimp from the Tortugas grounds alone amounted to $7.5 million in 1969.

There is a growing shrimp fishery in New England and the Alaskan area. Shrimp caught in these areas are smaller than the Gulf Coast shrimp.

778. What do oysters and clams eat? They eat microscopic plants called *phytoplankton,* which live in the water. Oysters and clams spend most of their time attached to or buried in the bottom and do not move around searching for food. They obtain food by pumping water through their bodies. Clams and oysters, like fish, breathe by means of gills, but their gills also act as a sieve, straining the phytoplankton out of the water. The gills are covered with a sticky substance which holds the phytoplankton. Tiny hairs, called cilia, move the food from the surface of the gills to the mouth. Pieces of food too large for the animal to eat are discarded and flushed out of the body by the water current. A single blue mussel will pump about 10 gallons of water a day, filtering out the plankton and debris that make up its diet.

779. Can oysters be fattened by supplementary feeding? Various nutrients have been tried experimentally as food supplements for oysters. Finely ground corn meal was found to be an economical method of producing fatter oysters.

780. Are oysters and clams always shucked by hand? Though there have been many attempts to develop automated devices to shuck oysters, clams, and other shellfish, none has ever proven to be sufficiently effective to be adopted and extensively used in the shellfish processing plants. In the decade just past (1960–70), fishery and engineering talents of several universities and laboratories have tackled this problem. They have tried vacuums, concentrated heat, electric shock, ultrasonic vibrations, and chemicals. Recent experiments using microwave energy show promising results. Oysters are exposed in a microwave oven just long enough to open them; when opened, extraction of the oyster is easy. There has been an increasing scarcity of capable oyster shuckers in recent years. This may solve that problem.

781. How do oysters produce pearls? Pearls are produced by a type of oyster more closely related to mussels than to the common edible oyster. These gems are formed by a foreign substance, such as a grain of sand, which accidentally enters the shell. The oyster attempts to reduce the irritation by depositing successive layers of nacreous material around the foreign body. The most valuable pearls are obtained from oysters taken near Ceylon and in the Persian Gulf. It is believed that the nucleus of these pearls is a tapeworm egg.

782. How are cultivated pearls produced?   Particles are inserted inside the shells of pearl oysters. It takes 3 to 5 years for an oyster to produce a pearl; if sea conditions are not right during the entire period, the oyster may yield only a worthless lump of calcium. Under the best conditions, only 60 percent of cultivated oysters live to yield pearls and only 2 to 3 percent produce gem-quality pearls.

783. How large is the Florida scallop fishery?   The calico scallop beds discovered off the Florida coast in 1960 have a potential yield of more than 300 million pounds per year. The beds extend from Daytona Beach to Fort Pierce and cover an area of 1,200 square miles.

784. Is it possible for lobsters to be grown and harvested commercially in private "ponds"?   The growing demand and increasing price of lobster meat in recent years makes lobster culture an extremely attractive venture for the 1970 decade. Records show that the United States has had to rely on Canada for more than one half of its annual lobster supply for the past 20 years. Based on experimental work completed through 1970, fishery specialists at U.S. and Canadian laboratories indicate that lobster culture in commercial ponds probably can be operated on a profit-making basis by using certain environmental factors to advantage. Warm water effluent from power-generating plants if impounded in ponds could be used advantageously; initial experiments using such ponds (temperatures 68° to 70° F) have cut in half the normal 5 to 7 years required to develop lobster of a commercially marketable size.

785. What is the usual reproduction rate of a female lobster?   Lobsters mate in the summer immediately after the female has molted and while the shell is still soft. The eggs which are dark green and about $\frac{1}{16}$ inch in diameter are attached under the tail by glue-like cement. They are carried for about 9 to 11 months. Hatching usually occurs in early summer and depends on the water temperature. The number of eggs laid depends on the lobster size; a 10-inch lobster will produce about 13,000 eggs, a 15-inch lobster about 49,000, and a 20-inch lobster about 124,000.

786. How often does a lobster molt?   When just hatched lobster larvae are about ⅓ inch long. They rise to the surface of the water to feed on plankton. In the next 15 to 35 days they almost double in length, and molt 3 times. At the end of this period they settle to the bottom and assume adult habits. As a juvenile the lobster will molt about 10 times the first year; molting becomes less frequent as the lobster grows older. A 1-pound specimen will molt once a year; a specimen weighing 10 pounds or more may not discard its shell for several years.

787. Why do lobsters and similar sea creatures molt?   Molting is nature's way of accommodating growth for lobsters and other crustaceans. Animals of the class Crustacea have no internal skeleton to serve as a framework for attachment of muscular tissues and organs. The chitinous exterior shell serves as a skeleton and also doubles as a protective shield for the animal. Crustaceans and other animals with exoskeletons must shed their shell if growth is to take place. Each time the lobster sheds his shell he will grow about 15 percent in length and 50 percent in weight. Molting, therefore, takes place more frequently in juveniles and smaller adults and also when more food is available.

XIII. MAN AND THE SEA

788. How deep has a skin diver gone? In December 1962, Hannes Keller, a Swiss mathematician, and Peter Small, a British journalist, descended to 1,000 feet in an open diving bell. At that depth, Keller swam outside for 3 minutes. He breathed a secret mixture of gases which was based on his own computations of what the human system requires and can tolerate. He also computed the decompression stages for the divers. Unfortunately, Small and another diver died during this attempt.

The decompression time of many days required for dives to 1,000 feet severely limits the economic value of diving to such depths.

789. How deep can a man dive without special breathing gases? The world record for deep diving on compressed air was set in 1968 by Hal Watts and A. J. Muns. According to *Skin Diver* magazine, Watts reached 390 feet and Muns stopped at 380 feet because of the onset of nitrogen narcosis. The dive was made near Government Cut off Miami Beach.

790. How deep can a man dive by holding his breath and using no supplemental breathing equipment? In February 1967, R. A. Croft, USN, established a world's record for a "breath-holding dive," reaching a depth of 212 feet and 7 inches. Later the same year during the week of December 16–22, the U.S. Submarine Medical Center of New London, Connecticut, organized a project off Ft. Lauderdale, Florida, for the expressed purpose of bettering that record. At that time, R. A. Croft set the present record depth of 217 feet, 6 inches. He made the descent in 40 seconds.

791. Were there any physiological effects noted during the deepest breath-holding dives? At the depth of 217½ feet, R. A. Croft reported some ear pain as the only noticeable physiological effect. However, physiological measurements taken during the dive showed some selective constriction of blood vessels to vital organs and an increased flow through these vessels as he went deeper. At the record depth, the pressure on his body was 111 lbs./sq. in. Croft has a remarkable lung capacity, 7.8 liters (1 liter = approximately 1 quart). The body at sea level is normally under 1 atmosphere of pressure, which is 14.7 lbs./sq. in.; the lung capacity of a person breathing normally is about 3 liters.

792. What is the usual time duration of a breath-holding dive for humans? According to those who have researched this subject extensively, most humans are limited to a breath-holding dive of less than a minute. Exceptions are professional pearl divers and some sports fishing skin divers who extend their underwater breath-holding dives to 2 or even three minutes by excessive hyperventilation.

793. How long can a diver work underwater? According to the U.S. Navy *Diving Manual,* a diver who works 2 hours at 100 feet must take another 2 hours and 12 minutes to come to the surface. For a 3-hour stay at 300 feet, over 19 hours would have to be spent in decompression. Divers living in underwater structures such as those tested by Cousteau and Link can work underwater for weeks at a time, because their blood becomes saturated with gases during the first 24 hours and the decompression time does not increase for longer underwater stays.

794. What advantages do hard-hat divers have over scuba divers? The main advantages are protective clothing and telephone connections to the surface. A hard-hat diver can work several hours at 200 feet, a depth beyond the reach of skin divers unless they have special diving apparatus and gas mixtures. Scuba divers, on the other hand, have the advantage of much greater mobility. Closed-circuit scuba systems now being tested are expected to provide safe breathing mixtures at depths exceeding 1,000 feet. These would provide the capability of working at great depths for extended periods.

795. What problems does a hard-hat diver have in going to great depths? First of all, he is tied to the surface by breathing tubes. Also, the high air pressure causes some of the nitrogen in the diver's air supply to dissolve in his blood (air is composed of about 80 percent nitrogen, 20 percent oxygen). The emergency limit for Navy divers is about 500 feet, although divers have gone much deeper.

796. What are some of the physiological dangers in deep diving?   Three serious problems that the deep sea diver faces are decompression sickness, nitrogen narcosis, and air embolism. *Decompression sickness,* better known as *bends* or *cassions disease,* is caused by absorption of nitrogen in the tissues. In order to work underwater, the diver's air supply must equal the water pressure. As the diver's body utilizes the oxygen from the air, the nitrogen remains in solution in the blood and tissues. If the diver does not return to the surface through the required decompression stages, the rapid change in external pressure creates bubbles of the gas in the blood or tissues, thus blocking circulation. The effects may be pain, paralysis, unconsciousness, and possible death.

Nitrogen narcosis: This condition is usually noted when breathing air at depths greater than 300 feet. At these depths where air is under high pressures, the diver's thought processes become unrelated. This seems to be a result of the high partial pressure of nitrogen, and at deeper depths can lead to unconsciousness or death. To avoid nitrogen narcosis, helium-oxygen mixtures have been substituted for air in deep diving.

Air embolism: This is the blocking of a diver's blood vessels by an air bubble. It is caused by excessive air pressure from the lungs leaking into the blood stream during a diver's ascent. Air embolism can occur if a skin diver holds his breath while ascending. In the extreme condition the lungs may rupture, in which case death is immediate.

797. What other problems result from diving?   Other severe problems are oxygen toxicity, carbon dioxide buildup in the lungs, and work limitation caused by increased gas density. Cold water chilling can also be a severe problem.

798. How long has scuba gear been in use?   In 1942, Jacques-Yves Cousteau and Émile Gagnan invented a regulator which would keep the diver's lungs at the same pressure as the surrounding water. In June 1943 their aqualung was successfully tested in the Mediterranean, off the French Riviera.

799. Why don't divers breathe pure oxygen?   Oxygen under pressure attacks the central nervous system. Symptoms of oxygen poisoning are twitching, dizziness, and nausea, followed by convulsions leading to death.

800. Why is helium used in breathing mixtures?　At depths of 100 feet, air becomes so dense that it is an effort for the diver to breathe. Below 300 feet, the effort required for breathing makes useful work impossible. Helium is substituted for nitrogen to make the mixture less dense.

801. What problems are caused by helium?　Helium alters the voice so that communication over electronic circuits becomes almost unintelligible; this problem becomes worse with increased depth. Because helium is lighter than air, it is a poor heat insulator, causing rapid loss of body heat for the diver working in cold water.

802. Can man learn to breathe water?　Death from drowning is caused by lack of oxygen. If water contains enough oxygen, men might be able to absorb oxygen through the lungs, just as fish absorb oxygen through their gills. At sea level, water will not hold enough dissolved oxygen to keep mammals alive, but at a pressure of eight atmospheres, enough oxygen can be added to the water to sustain life; this has been done with mammals in pressurized tanks. If man is to breathe water, some way will have to be found to dissolve additional oxygen in the water under pressure; there just isn't enough oxygen anywhere in the oceans to sustain him.

803. Where is the Graveyard of the Atlantic?　The title was applied to Cape Hatteras by Alexander Hamilton, who sailed by the area as a young man. Later, he used his influence as Secretary of the Treasury to have a lighthouse built at Cape Hatteras. The Cape is exposed to severe storm winds; it is open to the sea from north through east to southwest. Storms strike with sudden intensity. Hurricanes have driven many ships onto the beaches and shoals. The sands of Hatteras Island extend seaward as gigantic shoals for a distance of 12 miles; at some places they reach almost to the surface. Sand bars on Diamond Shoals are constantly shifting.

It is in this area that the southernmost portion of the Labrador Current meets the Gulf Stream. At times, the current has great velocity at Diamond Shoals; at other times there is no current or its direction may be reversed. With northerly and northeasterly winds, a dangerous cross sea is usually encountered.

804. What other areas claim this title?   Sable Island, a shifting area of sand off Nova Scotia, has claimed some 500 ships since 1800. The title "Graveyard of the Atlantic" or "Graveyard of ships" has also been applied to the Bahamas and other areas.

805. How many sunken ships are there?   During historical times an estimated one million ships have sunk. Even in peacetime, one of every hundred ships sinks each year. Only a few of those which sank near shore in shallow water have been salvaged; those in deep water are beyond salvage by present technology.

806. How are wrecks noted on navigational charts?   Charted wrecks are of two kinds: *stranded wreck,* where any portion of the hull is above the chart datum; and *sunken wreck,* where the hull is below the chart datum or when the masts only are visible.

Basic symbols used to designate wrecks are: a small silhouette figure diagram of a foundering ship with stern submerged is used to show a visible (or stranded) wreck; a horizontal line with three shorter lines drawn perpendicular to it and located at the midpoint and at an equal distance on either side is used to show a sunken wreck. A dashed or solid elliptical design with the word *wreck* printed beside or inside it, is used on large-scale charts. When wrecks or other obstructions are cleared by a wire drag, the maximum clearance depth is charted with a bracket under it.

807. What is the largest amount of gold that ever sank?   The ship carrying the greatest amount of gold to the bottom of the sea was not a Spanish galleon, but the British ship *Laurentic,* which was sunk by a German mine in January 1917. The ship sank in a depth of 132 feet off northern Ireland with 43 tons of gold worth $25 million. During seven years of salvage operations, all but 25 of the 3,211 bars were recovered.

808. Why isn't more gold salvaged?   There is no doubt that gold worth millions of dollars lies on the bottom of the ocean. It has been estimated that $150 million worth of treasure from Spanish ships which sank while crossing from the Caribbean to Spain has never been salvaged. The availability of scuba gear has opened the search beneath the sea to amateurs. Those who search in water less than 65 feet are almost certain to be disappointed; most treasure ships in these depths were located and salvaged soon after their loss. Waters between 65 and 200 feet deep, the effective working depth of scuba gear, offer most hope of finding treasure without expenditure of large capital. The hazards of salvage operations in deep water are great, and professional salvors must have a substantial margin of potential profits, because bad weather and equipment breakdown can make the operation expensive. Old wrecks are nearly always covered by coral, sand, and mud. Poor visibility adds to the difficulty of salvage operations.

809. Are many treasure hunters successful? There are fabulous true stories, such as the success of Wagner and Associates, who recovered more than a million dollars in treasure from Spanish ships off the coast of Florida. The story is told in *National Geographic Magazine,* January 1965. But for every successful treasure hunter, there are hundreds who don't even meet expenses. Perhaps there would be more success stories if it were not for the fact that successful treasure hunters are often closemouthed.

810. Are there sunken treasures more valuable than gold? The cargoes of thousands of ships sunk off the Atlantic coast of the United States during two world wars are more valuable than the gold of all the Spanish galleons. The cargo of one vessel is tin, reported to be worth $26 million.

811. How many ships are lost by stranding and collision? The year 1962 was exceptionally bad for strandings. In that year, 68 ships were lost, and damage was sustained by 925 more. Fourteen ships sank as a result of collisions and 1,804 ships were damaged.

812. Why hasn't the *Andrea Doria* been salvaged? Ever since the *Andrea Doria* sank in 1956 in 240 feet of water off the northeast coast of the United States, there has been considerable interest in salvaging her. She carried irreplaceable paintings by Rembrandt, which may still be undamaged. A life-sized bronze statue of Admiral Doria and a radar set in excellent condition have already been salvaged. Several salvage methods have been tried but have been unsuccessful because of nitrogen narcosis, bends, shark attacks, and foul weather. Further salvage attempts with a submersible vehicle are reportedly planned.

813. Why hasn't the *Lusitania* been salvaged?   In 1915 the *Lusitania* was sunk by a German submarine, coming to rest in 310 feet of water. She carried with her hundreds of tons of copper and bronze. After more than half a century, salvage attempts are still being planned.

814. Why are some Viking ships so well preserved?   Viking ships found in Danish waters have been in a remarkable state of preservation. The reason for this is that shipworms cannot survive in water of less than 7.0 ‰ (parts per thousand) salinity. In the Mediterranean and other areas of high salinity, timbers of sunken ships are quickly destroyed.

815. Has a concrete ship ever been built?   Yes. According to the best information available, 4 concrete experimental ships were built by the British during World War I. (An official New Jersey highway marker near Cape May, New Jersey, reports that 12 were built, but this could not be verified.) According to existing records, these ships proved a disappointment and were considered impractical after several trans-Atlantic crossings, because of their heavy construction. One of the ships was lost at sea, another was transported to Boston, a third is sunk south of Bimini off the Florida coast, and the fourth, known as the USS *Atlantis,* lies near the southern tip of the New Jersey coast in Delaware Bay where it is identified by a roadside "seamarker." It was purchased in 1926 and intended to be used with other hulls as a ferry dock. In the years since, storms have broken several portions of the hull and caused it to shift to its present angular position.

Another type of material known in the boat-building trade as *ferro-cement* is being used for boat construction. It utilizes cement as a binder and skin cover for a boat hull. It is finding increased use as boat-building material and is much thinner and far more suitable and durable in the water than older precast construction.

816. How are underwater artifacts located? Many artifacts are buried by mud and sand and are not visible to a diver. A special sonar, known as a *mud pinger,* has been used by Dr. Edwin Link to locate objects as much as 25 feet below the bottom. Magnetometers can locate iron and steel objects, but not bronze; underwater metal detectors of limited range are used to locate bronze art objects.

817. What Bronze Age artifacts have been found? George Bass of the University of Pennsylvania Museum has reported that the cargo of one Byzantine wreck was mostly copper ingots. Among the cargo was found a remarkably preserved wicker basket 3,000 years old. Stone hammers or mace heads and small bronze anvils have also been found on sunken ships.

818. What is the greatest depth from which it is possible to rescue survivors from a sunken submarine? Recent U.S. submarine disasters which involved the loss of the USS *Thresher* on April 10, 1963, and the USS *Scorpion* on May 21, 1968, had no survivors. Investigation of these sinkings indicates that the pressure hulls of both boats were destroyed before the bottom was reached. If the pressure hull of a submarine remains intact, escapes from boats on the bottom at a 600-foot depth are possible. In 1970 the British Navy successfully tested a new system in which 12 submariners rose to the surface in a simulated escape from a submarine, the *Osiris,* which was located at 600 feet. The escape was made through a standard escape tower; the men were wearing a newly invented, completely enclosed, hooded rubber suit. The escaping submariners breathed normally the air enclosed in the suit. They ascended at the rate of 10 feet per second. Other escape systems for use in much deeper water are under development by the U.S. Navy.

819. Has there ever been a successful rescue from a sunken submarine? Yes. Probably the most dramatic rescue on record is that of the 33 officers and men from the USS *Squalus.* The *Squalus* sank during a routine diving maneuver when an engine induction valve failed to close and the engine room was flooded. The date was

May 23, 1939; the location was off Portsmouth, New Hampshire, a few miles from the Isles of Shoals; the depth was 243 feet. The rescue was accomplished from the salvage ship USS *Falcon,* using a 10-ton two-compartment rescue bell designed by Charles Momsen and Allen McCann. This device was lowered over the escape hatch. The bell was secured to the submarine and the water was blown out with compressed air. Then the men in the upper compartment dropped onto the submarine to open the escape hatch and release those who were trapped. It required four trips in the *Momsen rescue bell* to recover the 33 men.

820. What is the Momsen Lung?   This submarine escape device was invented by Admiral Charles "Swede" Momsen in the late 1920s. It was the first successful escape equipment developed for submariners and was included as standard equipment on U.S. Navy boats after undergoing many tests. The first test, conducted on February 5, 1929, involved Charles B. Momsen and Chief Torpedoman Edward Kalinowski, who encountered no problems in surfacing from the Submarine S-4. The system was tested repeatedly down to as much as 200 feet with no failures. The Momsen Lung is strapped to the chest by the user. It resembles a gas mask with a rubber bladder and two tubes, one of which carries air from the mouth to the bag; the other tube encloses a canister with an air-purifying agent, soda lime, and carries the renewed air back to the mouth. The bag capacity is about equal in volume to the human lung.

821. Where may information on sunken ships be obtained?   During World War II the Hydrographic Office—now Naval Oceanographic Office—compiled a series of wreck charts and a "Wreck Information List" for the coastal waters of the United States. The Library of Congress, Washington, D.C. 20540, can provide photographic reproductions of these items, and, upon request, will furnish a price list of the items available. "A Descriptive List of Treasure Maps and Charts," compiled by Richard S. Ladd of the Library of Congress in 1964, is available for 30 cents from the Superintendent of Documents, U.S. Government Printing Office, Washington, D.C. 20402.

For information on shipwrecks from 1875 to 1940, the National

Archives maintains certain records and will make limited searches for documents, provided the inquirer furnishes the name of the ship and the date and place of the disaster. Address: National Archives and Records Service, Washington, D.C. 20408, Attention: NCRD.

A bibliography of popular books on sunken treasure may be obtained at no cost from the National Ocean Survey, National Oceanic and Atmospheric Administration, Rockville, Maryland 20852.

822. Where may information on salvage laws and tax laws be obtained?

Publications and pertinent data concerning sunken ships and salvage laws may be obtained from "Foul Anchor Archives," 25 Vale Place, Rye, New York, and also "Great Lakes Research, Inc.," 3274 East 93 Street, Cleveland, Ohio.

Information about tax laws related to wealth from sunken or buried treasure may be obtained from the Internal Revenue Service, 1111 Constitution Avenue, Washington, D.C. 20224.

823. When did man reach a half-mile depth?

In August 1934, William Beebe and Otis Barton reached a depth of 3,028 feet in the *Bathysphere* they developed. As early as 1930 they reached a depth of one-fourth mile and made a series of dives with the surface ship drifting and supporting the *Bathysphere* a few feet above the bottom. As he drifted along the bottom, Beebe telephoned instructions to the winch operator to raise or lower the sphere. This was the forerunner of the modern oceangraphic submersible.

824. How was Beebe's *Bathysphere* constructed?

The *Bathysphere* was designed and built by Captain John H. J. Butler and Otis Barton. This blue steel ball was 4 feet, 9 inches in diameter. The wall thickness was 1¼ inches and was designed to withstand pressures to 4,500 feet. It had a 14-inch hatch, two quartz portholes 6 inches in diameter, and rudders to keep it from spinning like a top when lowered on the ⅞-inch-thick steel cable. It was equipped with oxygen tanks, floodlights, and a telephone line to the surface but had no motors or engines or air conditioning. It had no fail-safe devices; if the cable holding the sphere had parted, no rescue would have been possible. The *Bathysphere* is kept and exhibited by the Smithsonian Institution in Washington, D.C.

825. How deep has man gone in the ocean?   On January 23, 1960, Jacques Piccard and Lieutenant Donald Walsh, USN, descended in the bathyscaph *Trieste* to a depth of 35,800 feet at the bottom of the Challenger Deep, the deepest known spot in the oceans. At that depth, the temperature was 2.4° C (36.5° F). During the descent they passed a minimum temperature of 1.4° C at 12,000 feet.

The *Trieste* was designed and constructed by Jacques's father, Auguste Piccard, the famous Swiss explorer of the stratosphere. It was launched in the Mediterranean in 1953 and acquired by the U.S. Navy in 1958.

826. How long did it take the *Trieste* to reach the bottom of the Marianas Trench on her record dive?   The dive with Dr. Jacques Piccard and U.S. Navy Lieutenant Donald Walsh aboard began at 8:23 on the morning of January 23, 1960. The logbooks indicate they report touching bottom at 1:06 in the afternoon. It took 4 hours and 43 minutes for the descent to the sea floor, 35,800 feet below.

Some dives by the *Archimedes,* the French submersible, in the Puerto Rico Trench report a much faster descent rate of 6 feet per second (360 feet per minute); but this is slow when compared to the world's fastest elevators located in the Pan American Building in New York City which operate at 1,700 feet per minute (5 times faster).

827. How do bathyscaphs like the *Trieste* operate?   The bathyscaphs of the 1960s differed from Beebe's in that they have no ties to the surface (Beebe's sphere was secured by a cable to a ship at the surface). The *Trieste* and similar submersibles might be called underwater balloons; they operate just as a balloon would in the atmosphere. The large spherical steel capsule that carries the crew and scientists can be compared to the balloon basket. The elongated craft to which the sphere is attached acts as a float and can be equated to the balloon; just as the balloon is filled with a gas that is lighter than air, so is this section of the submersible filled with gasoline which is lighter than water. This provides the lift necessary to return the submersible to the surface. Before beginning the dive, several tons of iron pellets are put in special compartments which can be opened;

the pellets are dropped out when the men want to rise from the bottom. Small battery-powered motors drive the propeller, steering, and other equipment which permits some maneuverability, but bathyscaphs like the *Trieste* are not designed for extended exploration on the bottom; these craft are engineered to be "elevators" to take men to the deepest depths and return.

828. When did Cousteau begin his diving saucer operations? Cousteau's first diving saucer, *Denise,* began operations in 1959. The saucer carried two persons at a speed of 1 knot, and had a depth capability of 1,000 feet. Water jets made it extremely maneuverable.

829. What was the first deep-diving submarine to reach a depth of a mile or more? The first deep-diving submarine, the *Alvin,* reached a depth of 6,000 feet on July 20, 1965, in a dive made in the Tongue of the Ocean area. William O. Rainnie, Jr., and Marvin J. McCamis were the crew for this dive which took place exactly 4 years before Armstrong and Aldrin walked on the moon—and, incidentally, there was no TV coverage of the event.

830. What are some of the physical characteristics of a deep-diving submarine like the *Alvin*? The *Alvin* is 22 feet long, has a beam of 8 feet, and displaces 13 tons. The basic hull structure of the *Alvin* is a sphere made of high-yield steel 1.33 inches thick and 7 feet in diameter. This steel was developed for use in nuclear submarines. The hull is close quarters for the crew of 2 just as a manned space ship is, and the deep-diving submarines utilize many of the miniaturized electronic devices and other products developed as a result of the space program to carry the needed instruments in their small capsule. The viewports of the *Alvin* are made of plexiglass, and floodlights on the exterior superstructure are placed to illuminate areas ahead of the ports. Three propellers drive the *Alvin,* a large one to move her ahead and two smaller ones used primarily to lift her.

Other deep-diving submarines like the *Deep Quest* made by Lockheed are of maraging steel, a tough steel originally developed for rocket cases. The *Aluminaut* of course is a special design made of aluminum.

831. Is the *Alvin* the same type of deep-sea vehicle as the *Trieste*? Both of these craft are deep-diving vehicles but they have quite different capabilities. The *Trieste* is essentially a bathyscaph as it is not designed for maneuverability. The *Trieste* operates as an elevator to take observers up and down in a spot; it cannot travel very far on the ocean bottom. Deep-diving craft such as the *Alvin, Deep Quest, Aluminaut, Deep Star,* and almost a dozen others have not only deep-diving capability but can also travel at speeds up to several knots while at depth or on the bottom.

832. What did *Aluminaut* prove? The Reynolds Metal Company submersible, *Aluminaut,* was built to prove the feasibility of aluminum for hull construction of a deep submersible. It proved itself in searching for the H-bombs lost off Spain and in the rescue of *Alvin,* the Woods Hole Oceanographic Institution submersible, which sank in water almost a mile deep. *Aluminaut* was designed to carry 6 men and operates at depths to 15,000 feet.

833. What is the longest submerged operation by a deep-diving vehicle? Research submarine dives are normally of short duration. Up to 1972 the longest uninterrupted dive was by the *Ben Franklin,* the Grumman-built submersible which drifted submerged at various depths to 1,800 feet in the Gulf Stream for 30 days. That mission started on July 14 and ended on August 14, 1969; the distance covered was 1,444 nautical miles, beginning off West Palm Beach, Florida, and ending off Cape Cod.

834. What was learned during the *Ben Franklin's* drift? Among the biggest surprises was the scarcity of fish in the Gulf Stream. This was attributed to the lack of food. The deep scattering layer was not found within the stream but was noted along its edges.

On the northern part of the drift, average speed was more than 3 knots; planning had been based on 1 to 1½ knots.

835. What are some of the specifications and performance characteristics of the *Ben Franklin*? Built by the Grumman Aerospace Corporation, the *Ben Franklin* was developed to serve as a deep submersible laboratory. The vehicle has an exterior length of almost 49 feet and a beam of slightly more than 13 feet; the height is

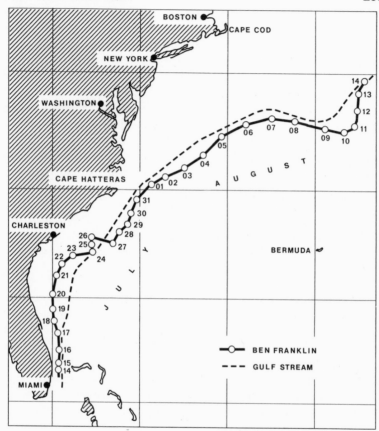

Drift of the Ben Franklin

20 feet. The interior pressure hull diameter is 10 feet. It displaces 138 tons, has a payload of 18,000 pounds, operational depth of 2,000 feet, and a collapse test depth of 5,700 feet. It can accommodate 12 persons for 4 weeks or longer and is capable of speeds up to 4 knots. There are 29 viewports for direct viewing or making a camera record of observations.

836. How deep can submarines operate safely? The maximum operating depth of submarines is a military secret; however, the engineering facts that determine safe operating depths are well known. The bathyscaph *Trieste,* which reached the deepest depth of the oceans, is no more like a true submarine than a stratosphere balloon is like an airplane. A true submersible should be positively buoyant and carry a considerable payload. A submarine built by today's methods to withstand a depth of 4,000 feet would not have sufficient buoyancy to carry a useful payload. Submersibles (not military submarines) have dived and operated under power at depths greater than 6,000 feet; *Alvin* and *Aluminaut* are two of these. *Aluminaut* has a depth capability of 15,000 feet. Newer construction materials, such as filament-wound, glass-reinforced plastic, produce high hull strength in respect to weight and may be used in the future for submersibles designed for greater depths.

World War I submarines had a capability of 100–200 feet; World War II submarines, 200–400 feet; and present-day submarines may reach depths of 1,500–4,000 feet. Small, high-speed interceptor submarines capable of diving to 6,000 feet or more are under development.

837. How do submarines navigate when submerged for weeks at a time?

When the nuclear submarine *Nautilus* made its famous voyage to the North Pole under the Arctic ice in 1958, the navigator was making use of Newton's second law of motion, $F = MA$ (force equals mass times acceleration). The navigation system, known as inertial navigation, uses accelerometers to continuously sense changes in velocity with respect to a known starting point. Three gyroscopes, one for each direction of movement, create a platform which remains stabilized regardless of maneuvers of the submarine. The system is entirely independent of magnetic influences; this is an essential requirement in polar navigation.

In addition to inertial navigation systems, submarines may rely on acoustic positioning sources on the bottom of the ocean to locate known points of reference, and they can make use of Doppler sonar to determine ground speed. The whole Doppler-inertial navigation system on a nuclear submarine is tied together by an electronic computer.

838. What was the first undersea habitation?

In 1962, Cousteau carried out the first successful experiment in underwater living. The underwater structure, known as *Conshelf One,* was lowered to the bottom in 33 feet of water off Marseille and occupied by two men for a week.

839. What experiments has the United States conducted in underwater living? The United States program was pioneered by Captain George F. Bond, a medical officer in the United States Navy, who discovered that, once a diver's blood has become saturated with breathing gases at a given depth, decompression time is related only to the depth and not to the length of time the diver remains there. Dr. Edwin Link, inventor of the *Link Trainer,* was one of the pioneers in the U.S. undersea habitat program. Projects have been carried out on both the Atlantic and Pacific sides of the United States. Off California, one diver remained underwater 45 days in *Sealab.* Commander Scott Carpenter was one of the participants in this project. In the Virgin Islands, project *Tektite* was managed by the Navy with Department of Interior, NASA, and industrial support.

840. What other countries have conducted underwater living tests? The Soviet manned-diver habitat program began in 1965; at least seven structures have been placed in shallow water. The Germans have experimented with a new bottom habitat to conduct biological studies off the island of Helgoland in the North Sea. The British are constructing shallow-water habitats; the Canadians have made tests in a habitat at 50 feet in Lake Erie, and the Czechs, Bulgarians, Poles, East Germans, and Cubans have all experimented with habitats in shallow water.

841. What is the approximate amount of minerals in a cubic mile of seawater?

Sodium chloride (common salt) . . .	128,000,000 tons
Magnesium chloride	17,900,000 tons
Magnesium sulfate	7,800,000 tons
Calcium sulfate	5,900,000 tons
Potassium sulfate	4,000,000 tons
Calcium carbonate (lime)	578,832 tons
Magnesium bromide	350,000 tons
Bromine	300,000 tons
Strontium	60,000 tons

Boron . . . .   21,000 tons
Fluorine . . . .   6,400 tons
Barium . . . .   900 tons
Iodine . . . .   100 to 1200 tons
Arsenic . . . .   50 to 350 tons
Rubidium . . . .   200 tons
Silver . . . . ,   up to 45 tons
Copper, lead, manganese, zinc . . . . . .   10 to 30 tons
Gold . . . .   up to 25 tons
Uranium . . . .   7 tons

842. What minerals in seawater are commercially valuable? Only sodium chloride, magnesium metal and some of its compounds, and bromine are extracted directly from seawater in large quantities. Several calcium and potassium compounds are obtained as by-products in the extraction of other minerals.

843. Why aren't more minerals recovered from the sea? Seawater is an extremely low-grade source for most of the 60 naturally occurring elements found in it. The important industrial metals such as aluminum, tin, copper, zinc, silver, iron, nickel, and manganese are valued at less than $1.05 per million gallons of seawater, and it has been estimated that the $20,000 worth of gold in a cubic mile in the ocean would cost far more than its value to extract.

844. What other minerals might be extracted from seawater in the future? If processing costs could be reduced, strontium, rubidium, lithium, and fluorine might be extracted. If extraction is combined with desalting of water in nuclear-powered plants, boron, uranium, copper, and manganese might be recovered. It is likely that research will be concentrated on elements which are in short supply, such as copper.

845. How is bromine obtained from seawater? In the United States, bromine has been obtained commercially from the sea since 1933. Seawater is treated with sulfuric acid and chlorine is added; bromine is released as a vapor and is combined with an alkaline solution. About half the bromine used in the United States, and 80 percent of the world's bromine supply, comes from the sea. The supply is inexhaustible; 1 cubic mile of seawater could supply all the bromine needed by U.S. gasoline refineries for two years.

846. How much of the U.S. supply of magnesium comes from the sea? In January 1941 the Dow Chemical Company began extracting magnesium from seawater at Freeport, Texas. Since then, other companies have begun producing magnesium hydroxide from the sea, and the entire U.S. supply now comes from the oceans. Although seawater does not contain more than 0.13 percent magnesium, the process is cheaper than mining and refining ore.

847. How is magnesium removed from seawater? Seawater is treated with lime, precipitating magnesium hydroxide, commonly known as *milk of magnesia*. This is treated with hydrochloric acid and evaporated, leaving a residue of magnesium chloride. Metallic magnesium is obtained by passing an electric current through the magnesium chloride.

848. How is iodine concentrated from seawater? The 1 gram of iodine in 20 tons of seawater is concentrated into 200 grams of dry kelp. The dried seaweed is burned to obtain iodine. Even the iodine obtained from the salt beds of Chile comes from the remains of kelp.

849. How much of the world's salt is taken from the sea? About 35 million tons of salt are obtained every year from evaporation of seawater; this is about one third of the total world production from all sources. Most of this is used for snow removal, water softening, and refrigeration.

850. Is there an efficient method of obtaining fresh water from the sea? Compared to the cost of purifying fresh water, desalination is not yet an efficient method of obtaining water for either drinking or irrigation. Only in a few water-poor areas is it now economical to desalinate seawater. As commercially practical plants reduce water costs, the consumption of water will increase, making the desalination operation not only attractive, but also essential.

Use of water in the United States is increasing at the rate of 1.5 million gallons an hour! In many parts of the country, water shortages are already critical. The best commercial methods available in 1952 produced 1,000 gallons of fresh water from seawater at a cost in excess of 4 dollars; today the cost is about 1 dollar. Future costs of 20 to 30 cents are possible.

851. Why have governments become so interested in the development of commercial water desalting techniques since 1950? In 1900 the water requirement in the United States was about 40 billion gallons per day (bgd); by 1920 it more than doubled (91.5 bgd) and in 1940 reached 136.4 billion gallons per day. Figures for 1960 and 1965 are 323 billion and 371.1 billion gallons per day, respectively. It is estimated that by 1980, the water use rate in the United States will reach 500 billion gallons per day and by the year 2000, a trillion gallons per day. Development of economically efficient industrial methods for supplying a significant portion of the water needs from the limitless ocean resources is therefore considered essential if the future populations of the world are to have water in sufficient quantities.

852. What are the methods by which fresh water is produced from seawater? Three basic techniques are used to produce fresh water from saline. These are: distillation, a conversion process which employs the basic concept of boiling the salt water and condensing the resulting steam; use of a membrane or a physical barrier which allows the separation of salt water by preferentially permitting the passage of salt while restricting the passage of water or vice versa; and crystallization which employs use of hydrates or freezing to separate the water and salt. About 98 percent of all plants in operation utilize the distillation technique; the membrane and crystallization systems are used in the remaining 2 percent of the facilities.

853. What is the world total of fresh water produced by desalination methods? As of January 1, 1969, the world fresh-water output from desalination plants totaled 247,166,000 gallons per day. Based on the worldwide historical development of desalination plant capacities, it is estimated that the output by 1975 will reach one and a quarter billion gallons per day.

854. What is the total number of seawater desalination plants in operation? The Department of Interior's Office of Saline Water

Desalting Plant Inventory listed a total of 686 desalting plants with a 25,000 gallon per day capacity or greater in operation or under construction as of January 1, 1969. Fifty-four of these plants represent new construction begun in 1968.

855. Geographically, what is the present distribution of desalting plants? The Middle East leads the world in total desalting capacity; 74 plants supply 62.9 million gallons per day (mgd) of fresh water. The first large plants producing a million gallons per day or more were installed in the oil-producing desert areas surrounding the Persian Gulf. The United States and its territories produce 53.4 mgd of fresh water from 322 facilities. Russia has 7 plants with a productivity of 37.2 mgd. There are 88 plants in Europe providing 29.6 mgd. In the Caribbean area 18.5 mgd is the output from 26 plants; in England and Ireland 15.9 mgd from 63 plants; in North America (except U.S.) 8.5 mgd from 12 plants; in South America 3.8 mgd from 21 plants; in Asia 3.2 mgd from 24 plants and in Australia 1.3 mgd from 6 plants.

856. Where is the largest water desalination plant and what is its capacity? The desalination plant located at Rosanta, Mexico, produces 7.5 million gallons per day (mgd) and is the largest facility in the Americas. One of the largest European plants, located at Terneuzen, Netherlands, provides 7.6 mgd. The largest Russian plant is situated at Schevelenko. It is planned to have the ultimate capacity of 31.7 mgd, and probably deserves to be identified as the world's largest such facility. The largest plant in the United States supplies 2.6 mgd of fresh water to Key West, Florida.

857. How many water desalination plants are in operation in the United States? There were 307 water desalination plants operating in the United States by 1969, another 15 were operating in five U.S. Territorial Possessions. Thirty-seven of the 50 states have plants producing fresh water at a rate of 25,000 gallons or more per day. Desalting plants are operating in 69 other countries.

858. What is the primary use of fresh water produced by desalination? According to the United Nations, cities are the main users of water-desalting plants. About 50 percent of the world's water desalting capacity is used to supply municipal needs. Industrial demands for fresh water almost equal municipal requirements, taking 48 percent of the world capacity. The remaining 2 percent is used to satisfy military, tourist, and special domestic needs.

859. Is snow or seawater the source of water for the U.S. Antarctic Scientific Stations? To obtain water at most of the stations means shoveling snow and melting it. And because fuel to operate the melters needs to be flown from McMurdo to the inland stations, water is used sparingly. The water supply at McMurdo, however, is produced by desalting seawater. A nuclear reactor provides the heat needed to operate the desalination system which uses seawater pumped from McMurdo Sound. The desalting is done in two 16-stage flash evaporators, each capable of producing 14,000 gallons of fresh water daily. Approximately 10 gallons of seawater must be processed to produce 1 gallon of fresh water. Electrical heating tapes are used to keep the distribution pipes from freezing.

860. How valuable is the gold in seawater? Although analyses of seawater samples from various parts of the ocean differ only slightly when a volume of one cubic yard of seawater is considered, extrapolations for the amount of gold in a cubic mile magnify substantially the analytical differences and geographic variations. Estimates vary from $20 thousand to $93 million for a cubic mile.

After the First World War, Germany seriously considered extracting gold from the ocean to pay the war debt. The idea was endorsed and supported by the distinguished chemist Fritz Haber. One of the main goals of an extended series of expeditions by the *Meteor,* which crossed and recrossed the North and South Atlantic repeatedly between 1924 and 1928, was to investigate the feasibility of gold recovery from the ocean. Although the quantity of gold found was less than expected and the cost of extraction prohibitively high, the *Meteor* collected much valuable oceanographic data.

861. What sea floor minerals are now exploited? Oil and gas are by far the most important; their value is many times that of all other mineral resources combined. The annual production is in excess of $6 billion. Sand and gravel, oyster shell, tin, gold and other heavy minerals are dredged from the sea floor. Sulfur is recovered from beneath the sea bottom by melting it with superheated water and piping it to the surface. Coal, iron ore, and nickel-copper ores are mined from beneath the sea in shafts beginning on land. Coal from beneath the sea accounts for more than 30 percent of Japan's production and more than 10 percent of Britain's.

862. What minerals on the deep-sea floor are of economic interest? Although minerals on the continental shelf are the only ones that can be profitably recovered at present, it may be necessary to exploit deep-sea deposits in the future as land resources are exhausted. Known materials include manganese nodules, an ore of manganese, copper, zinc, and other metals; phosphorite, a source of phosphate fertilizer; red clay, containing aluminum and copper; globigerina ooze, which could be used for cement lime; diatomaceous ooze, a source of silica; and barium spherules.

863. When were manganese nodules discovered on the sea floor? In 1875, HMS *Challenger* dredged manganese nodules from the ocean bottom during her epic 3-year round-the-world cruise. Nevertheless, it was not until the International Geophysical Year in 1958 that systematic studies of distribution were begun.

864. How are manganese nodules formed? The black concretions of iron-manganese are believed to have been formed by precipitation from seawater. It is still not known whether they are formed by a purely chemical process or whether bacterial activity is also involved. Some scientists believe that submarine volcanic emanations play a part in the nodule formation; others disagree. One thing is known—they form very slowly; radioactive dating indicates that less than a millimeter of thickness is added in a thousand years.

865. How abundant are manganese nodules? Reserves have been estimated to be 140,000 to 200,000 tons per square mile over areas of more than 16 million square miles. The supply is inexhaustible; even though the rate of formation is extremely slow, they are forming faster than the world consumes manganese.

Photographs of the deep-ocean floor have shown parts of the Pacific Ocean bottom to be 20 to 50 percent covered.

866. Are manganese nodules now being dredged commercially? Only on a limited basis. One vessel, *Deepsea Miner,* dredged 1,600 tons per day in waters 3,000 feet deep during trials. At that rate, one ship could supply 16 percent of the daily U.S. need for manganese and 32 percent of the cobalt and nickel. Of course the nodules are not all located in 3,000 feet of water; most are at depths of 12,000 to 20,000 feet.

867. How is phosphorite formed? It is uncertain whether the formation of these precipitates is entirely an inorganic process or whether biological activity is involved. Phosphorite may be found as nodules, grains, or slabs. Deposits are generally found in water from 300 to 1,000 feet deep.

868. How abundant is phosphorite? The most promising deposits near the United States are off southern California, amounting to 50 or 60 million tons. The phosphorite contains about 30 percent phosphate and could be marketed as fertilizer in the west coast states or shipped to Asiatic countries bordering the Pacific.

Other deposits are known off the southeastern United States, Republic of South Africa, northwestern Africa, and western Australia. Because of the large supply of phosphate deposits on land, these marine deposits are not economically interesting at present.

869. How is sulfur mined from beneath the sea? Oil drillers discovered that sulfur was associated with some salt domes and could be recovered cheaply by pumping superheated water down to melt the sulfur, which is then forced to the surface by compressed air. About 20 salt domes have been mined successfully under the land; almost half of them are now exhausted, making it necessary to search for sulfur under the sea, which can be recovered by the same technique used on land.

870. When was sulfur first mined beneath the sea? The Grand Isle Project, 7 miles off the coast of Louisiana, began operation in June 1970; this was the world's first offshore sulfur mine. A platform was built in 43 feet of water to support the drilling rig. The sulfur deposit varies from 220 to 425 feet in thickness and covers several hundred acres. Two mines off Louisiana account for 15 percent of the U.S. production of sulfur.

871. Is gold mined from the bottom of the sea?   Most gold in ocean-bottom sediments comes from lodes on land. The cost of mining and concentrating gold offshore is considerably greater than onshore, so only a large, high-grade deposit would justify commercial operations. Exploration has been conducted in several areas. Beach sands along the Nome, Alaska, Gold Coast yielded $100 million worth of gold during the gold rush; these same sands are known to extend beneath the ocean. Exploration is being conducted along the coasts of North Carolina and Oregon to determine whether gold can be dredged economically in these areas. The Soviet Union has discovered gold-bearing sands at a depth of 82 feet off the Siberian coast and reportedly plans to mine the deposits.

872. Where are diamonds mined from the seabed?   Off the southwest coast of Africa more than one-half million carats of diamonds have been dredged from the sea floor since 1952. The ship *Rockeater,* owned by a United States company, dredges about $200,000 worth of diamonds a month from depths as great as 200 feet. The diamonds are found in mixtures of sand, gravel, and boulders. Although the diamonds can be separated easily, dredging techniques which can handle large quantities of sand economically must be developed before the operation will be profitable.

873. Are there valuable deposits of tin on the sea floor?   About 75 percent of the world's tin comes from stream deposits, but tin cannot be transported long distances in water without breaking down. Because of this, only areas near stream mouths are potential sites of marine placer deposits. About 1 percent of the world's tin comes from marine deposits. Production could probably be doubled by mining additional submerged former stream channels off Indonesia.

874. What mineral resource is dredged in the greatest quantity?   Sand and gravel are mined from the sea bottom in greater quantities than any other material; more than half a billion tons are produced annually in the United States. Most of the sand and gravel is used in concrete. Small amounts of sand are used in glassmaking and abrasives. As land deposits become exhausted, the sea will become the major source of sand and gravel for the construction industries, particularly in coastal areas.

875. What are the commercial uses of shells?   Shells are the basis of a major industry in Faxa Bay, Iceland. Shells broken by winter storms are swept into the bay by tides. The concentrated deposits are dredged up for use in the manufacture of cement. Deposits are increasing faster than they can be dredged.

Shells are also used for the manufacture of lime, high-quality buttons, and road metal.

876. What is glauconite?   This greenish mineral is a hydrated potassium, iron, aluminum silicate found on the continental shelves of many parts of the world. Glauconite could be used as a source of potash in fertilizers or in the manufacture of potassium compounds.

877. How much of the world's oil and gas comes from beneath the sea?   About 6½ million barrels of oil are produced every day from subsea wells. This represents about 16 percent of the world's production of oil. About 6 percent of the natural gas comes from offshore wells. By 1980, offshore production of oil is expected to be one third of total world production.

878. Where are the major oil-producing regions?   Areas producing at least a million barrels a day include the Persian Gulf, Lake Maracaibo, and the Gulf of Mexico off Louisiana. Other areas of significant production are off California, Alaska, the North Sea, South China Sea, and Caspian Sea.

More than 50 companies operated more than 7,000 wells in the Gulf of Mexico in 1970, according to Dr. W. T. Pecora, director of the U.S. Geological Survey. Some 16,000 oil wells have already been drilled off U.S. coasts and the number is increasing by 1,000 each year.

879. How great are the continental shelf oil reserves?   Nobody knows for sure. Some geologists estimate that 40 percent of the earth's oil reserves lie beneath the continental shelf, but it is possible that subsea oil deposits exceed those on land.

880. When were offshore oil wells first drilled?   Oil wells were drilled from piers along the coast of southern California in 1891. Drilling began in the Gulf of Mexico about 1940.

881. Is oil now being formed on the ocean bottom?   Dr. Claude ZoBell of Scripps Institution of Oceanography has made exhaustive studies of the role of bacteria in the formation of petroleum. He estimates that oil is now forming in sediments as fast as it is being removed by drilling. The oil is, however, too widely diffused to be recovered.

882. Is it possible that there are other Arctic Ocean oil deposits equal in size to the Alaskan northern slope discovery at Prudhoe Bay?   According to reports released by USSR's scientists and information appearing in the Soviet newspaper *Pravda,* exploration for oil in the USSR Arctic region has been under way for some time. Mikhail Kalinko, division chief of the All-Union Scientific Research Institute for Petroleum Geology Exploration, indicated that initial scientific studies of the Soviet Arctic indicate a similarity in geological structure between many northern areas and the Gulf of Mexico oil-bearing formations. He also likens the Mezenskaya Basin adjoining the White Sea to the Northern Alaskan Basin, implying that it offers good possibilities for oil finds. Though no oil finds in the Arctic have been reported by the USSR up to mid-1970, there are several locations that offer good indications of productive fields. These are mainly on or near the Arctic Islands of Spitsbergen, Severnaya Zemlya, and the Novosibirskiye Islands. Several large gas fields in the Arctic Zone near the sea have already been tapped; these are located near the Pechora Sea, the Ob Gulf, and the outlet of the Taz River; the Urengeiskoye find located in the mid-Siberian basin midway between the Ob and Taz Rivers is claimed to be the world's largest gas field with a reserve of 3 trillion cubic meters.

883. What is the greatest depth of water in which ocean-floor drilling can be accomplished?   The *Glomar Challenger* is a specially constructed platform designed specifically for the purpose of drilling into the ocean floor. It is under the technical control of marine geologists of the Scripps Institution of Oceanography and is exclusively dedicated to work on the Deep Sea Drilling Project. She is at sea almost continuously taking cores from the ocean in various depths of water. The first core ever obtained by the *Glomar Challenger* was taken on August 13, 1968, from the Sigsbee Plain in the Gulf of Mexico; there the drill penetrated the bottom to 2,528 feet in 9,259 feet of water. Later the same month, drilling in a depth at 12,327 feet was accomplished by taking a core of 2,060 feet. Subsequently a successful core was taken from the Puerto Rico Trench near San Salvador in almost 17,500 feet of water. Cores from deeper depths will probably continue to be collected by the *Glomar Challenger* as scientists probe the ocean bottom to obtain more scientific facts about the geologic past.

884. Is farming the sea practical?   It is questionable whether environmental conditions in the open sea can be modified to increase productivity, although many imaginative ideas have been proposed.

At the present time, shallow enclosed areas offer the greatest hope of economic return. The cultivated species are pearl oysters or expensive fish and shrimp.

Japan has developed fish farming and aquaculture to a higher degree than any other country. Fish-farming centers have been established in the Inland Sea to offset the decrease in catch of high-quality fish in coastal waters. Eggs are hatched and fries released into waters of the Inland Sea.

885. What are some of the problems of aquaculture?   Fertilizers used in enclosed areas of the sea have stimulated growth of weeds and unwanted species as well as of desirable fish. These areas are sometimes overfertilized from nutrients in sewage discharge and from detergents.

The coastal areas which are best adapted to aquaculture are also the most polluted. One form of pollution, thermal pollution, may be used to advantage by transplanting heat-tolerant marine plants to waters warmed from power plants.

886. What is an artificial reef?   An artificial reef is a shoal area which has been constructed for the purpose of giving additional hard-surfaced irregular topography to the ocean bottom. This provides appropriate surfaces for attachment of marine growth and in turn serves as a feeding area and habitat for fishes. Such reefs are usually established near resort communities and within a few miles (generally less than 25) of the shore. Once established, they are favored sites for sport fishing and charter boat parties.

887. Do artificial reefs increase fish production?   Most artificial reefs have been built for sport fishermen rather than for commercial fishermen. Fish catches have increased dramatically for a few years and then have sometimes decreased, possibly because of chemical changes in the water brought about by corrosion.

888. Where was the first artificial reef established along U.S. shores?   The first such reef was constructed in 1935 at Cape May, New Jersey, by a group of avid saltwater sport fishermen.

The following year a second reef was built by the Reading Railroad off the resort community of Atlantic City, New Jersey. This was strictly a business venture; after it was established, the railroad advertised a "fisherman's excursion" for $1.50, which included a round-trip ticket from New York and a full day of fishing, including boat and bait.

889. How are artificial reefs constructed?   Once a desirable location for an artificial reef is selected, little additional preparation is required other than collection of the selected solid waste materials needed for the reef and transportation (barging) to the site. Solid waste in this particular connotation means scrap, junk cars, broken concrete slabs, culverts, cement blocks and weighted building materials, old tires, junk steel and old ships, barges and boats. All materials put down for the reef must be weighted, particularly the tires. In some cases, currents and wave action have caused junk cars to shift position until "sanded in"; therefore, the practice of cabling several cars together is recommended. Often, tires are put together in various designs and arrangements to provide maximum surfaces and crevices for marine-life attachment and fish habitats.

890. Can artificial reefs be built anywhere?   Usually the sites selected are on relatively smooth beaches of the continental shelf not too far offshore and near resort communities with marinas and sport fishing and charter boat fleets. The purpose is to provide additional environmental habitat and feeding areas for various fishes sought by sport fishermen. Irregular-shaped, heavy, hard scrap materials that stay on the bottom provide the attachment base for marine growth that serve as fish food and the crevices for hiding and protection.

Artificial reefs cannot be built in U.S. waters without official permission from the Navy, because they can shelter submarines from sonar detection.

891. Can raising fish in hatcheries increase populations? Perhaps, if the fries are sheltered beyond the time of maximum danger, a population increase might result. Adding newly hatched fish to the ocean would have almost no effect, because of the high mortality rate during the first few weeks after hatching.

892. What fish species might be profitably farmed? Research is being conducted with a number of high-value species including pompano, salmon, and redfish. Fish might be combined with other organisms, such as oysters, scallops, or lobsters, to obtain a greater return per acre.

893. Why are shrimp cultivated in Japan but not in the United States? This is a matter of economics. In Japan the price per pound of shrimp varies from 2 to 3 dollars and 50 cents, depending on the season. In times of shortage, the price may go over 4 dollars. In the United States the price is much less. Large quantities of shrimp have been discovered in recent years in the Dry Tortugas and the supply has kept the price down.

White shrimp have been grown experimentally in a pond by the Galveston Laboratory of the National Marine Fisheries Service. Shrimp one-half inch long have been raised to the 4½-inch size in 5 weeks. Production as high as 600 to 800 pounds per acre can be attained.

894. How can oyster production be increased? The Japanese have been growing oysters on ropes suspended from rafts for forty years. Advantages of this method are that nutrients in the water from surface to bottom can be used and the oysters are removed from some bottom predators. The raft method results in yields of more than 30,000 pounds of oysters per acre; the old method yielded no more than 600 pounds per acre annually.

Oyster culture is also highly developed in the Mediterranean Sea, where oysters are harvested from sticks thrust into the shoal bottom.

895. Can marine plants be farmed? Culture of edible seaweed species is practiced widely in Japan. Experimental work in California indicates the possibility of improving kelp beds by control of predatory sea urchins.

Hydrocolloids from seaweed have made possible many convenience foods and have served as homogenizing agents in toothpastes, pharmaceuticals, and cosmetics. Industrial applications include ink, paint, and tire production. The use of marine colloids has become so widespread that the raw material is in short supply and the artificial culture of highly productive seaweeds is being considered.

896. How many people live near the ocean? In the continental United States, 52 million people live within 50 miles of the coastline. Although this coastal belt contains only 8 percent of the land of the 48 contiguous states, it is occupied by 29 percent of the population. Most of the population of Hawaii and Alaska is within 50 miles of the ocean.

The states bordering on the oceans and Great Lakes contain 75 percent of the U.S. population. Millions more visit the seashore for swimming, boating, fishing, and relaxation.

897. How many boats are used for recreation in American waters? The Coast Guard estimates that more than 20 million people participate in recreational boating in coastal waters and the yearly increase in boats used in American waters is more than 200,-000. In 1969, expenditures for recreational boating amounted to $3 billion.

898. What are some of the law-enforcement responsibilities of the Coast Guard? The tremendous increase in the number of pleasure boats on navigable coastal and other U.S. waters in the past few years has substantially increased the Coast Guard's contacts with the public.

The Coast Guard can impose a civil penalty for failure to comply with the numbering requirement, failure to observe the Rules of the Road (the greatest cause of collision), failure to report a boating accident and similar violations. Most violations subject the owner and/or operator to a statutory penalty. Reckless or negligent operation of a vessel which endangers life, limb, or property is prohibited by law. A civil penalty for negligent operation can be imposed; as much as $2,000 and up to one year imprisonment may be levied if the operator is convicted of a criminal offense of reckless and negligent operation.

When hailed by a Coast Guard vessel or patrol boat, one is required to stop immediately and lay to or maneuver in such a way as to permit the boarding officer to come aboard. Failure to stop or permit boarding may subject the operator (or owner) to a fine of $100. Coast Guard boarding vessels are identified by the Coast Guard ensign (flag) and uniformed personnel. The Coast Guard Auxiliary offers a free courtesy examination to boat owners who desire to have their craft checked for boating safety and all other legal requirements. If there are irregularities, no report is made to any law-enforcement authority. The Auxiliary examiner advises the owner of all deficiencies so that they can be corrected.

899. Why is there a number on the bow of most small pleasure boats? The Federal Boating Act of 1958 established a uniform system for identification of pleasure craft. Undocumented vessels are numbered in the state in which principally used. If powered by equipment over 10 hp and used on navigable waters of the United States in New Hampshire, Washington, Alaska, District of Columbia, Guam, or Puerto Rico, the *Certificate of Number* is issued by the Coast Guard. The identification number issued to a vessel is shown on the certificate which must be on board the vessel when in operation. The same number is painted on or attached to each side of the bow. The numbers must be block characters of a color contrasting to the background and at least 3 inches high. Fees are charged by the states and by the Coast Guard for localities mentioned previously for issuing Certificates of Number. Fees are fixed by state law; the Coast Guard numbering fee is 3 dollars and is valid for three years. Upon sale or transfer of a vessel, when it is to be used in the same state, the old number will be issued to the new owner. Numbers cannot be transferred from one boat to another except in the case of dealers or manufacturers who obtain special permission. These numbers provide a means of identification for the many small pleasure craft operating in navigable waters. The Coast Guard is responsible for safety, rescue, and police enforcement of marine regulations and laws. The numbering system is in essence a license much like the automobile license which states issue and is essential for identification of craft and their owners.

900. How many Americans participate in salt-water sport fishing?   According to the Bureau of Sport Fisheries and Wildlife, 8.3 million anglers spent $800 million to catch 1.7 million tons of salt-water fish in 1965.

901. How many scuba divers are there?   About 2 million active scuba divers in the United States invest $40 to $50 million annually in equipment and considerably more in the sport itself. Each year between 50,000 and 100,000 people begin scuba diving. There were 1,800 diving clubs and national societies in the United States in 1969 with an average of 20 active members per club. The great majority of recreational scuba divers are not affiliated with clubs.

902. How rapid is the growth in coastal recreation activities?   The table below was compiled from statistics and projections of the Departments of Commerce, Interior and Transportation:

Type of Recreation	Participants (millions)	
	1964	1975
Swimming	33	40
Surfing	1	4
Skin diving	1	3
Pleasure boating	20	50
Sport fishing	8	16
National Park and Forest recreation . . .	18	44

XIV. POLLUTION

903. How does the ocean become polluted? As the writer of *Ecclesiastes* knew, all the rivers run into the sea; these rivers carry more oil to the sea than all the tanker oil spills combined. They also carry sewage, agricultural fertilizers, pesticides, mill wastes, detergents, radioactive wastes, and many other pollutants. Most of the pollution of the oceans takes place at the margins, where man's activities are concentrated.

Activities such as dredging and construction of jetties alter natural processes and cause pollution. Heating of coastal waters by industrial discharge decreases oxygen-carrying capacity of the water, affecting marine life.

904. How much waste is dumped into the sea each year? In 1968 about 48 million tons of wastes were dumped into the sea. This included dredge spoils, industrial wastes, sewage sludge, construction and demolition debris, solid waste, explosives, chemical munitions, radioactive wastes, and miscellaneous materials.

905. What is the greatest pollution to U.S. waters? Sediments carried by erosion represent the greatest volume of "wastes" entering surface waters. Pollution specialists estimate that silt and suspended solids reaching streams, rivers, lakes, and estuaries of this country are at least 700 times greater than the total sewage discharge loadings. Erosion can occur on cropland, unprotected forest soil, overgrazed pasture, strip mines, roads, and bulldozed urban developments. According to experts, erosion at highway construction sites in an average rainstorm is 10 times that of cultivated land, 200 times that of pasture land and 2,000 times that of a forest area.

906. What are the total tonnages and the relative proportions of wastes dumped off the shores of the United States? Figures in the following table are taken from the 1970 report to the President of the United States, prepared by the Council on Environmental Quality. Ocean dumping for 1968 is listed by types and amount:

Waste type	Total (in tons)	Percentage of total
Dredge spoils	38,428,000	80
Industrial wastes	4,690,500	10
Sewage sludge	4,477,000	9
Construction & demolition debris	574,000	1
Solid wastes	26,000	1
Explosives	15,200	1
Radioactive wastes	Tonnage insignificant; ton-	
Chemical munitions	nage does not provide ac-	
	curate indication of pollu-	
	tion potential.	
Totals	48,210,700	100%

907. What is the greatest source of man-made pollution of U.S. coastal and estuarine waters? A Federal study issued in 1970 reported that industrial liquid wastes are the largest source of pollution for coastal and estuarine areas. Municipal liquid wastes are second, followed by agricultural pollutants, which include animal wastes, pesticides, and fertilizers.

908. Of all pollutants dumped in the ocean, which accounts for the largest amount? According to the 1970 report issued by the Council on Environmental Quality, dredge spoils account for 80 percent by weight of all ocean dumping. The Corps of Engineers estimates that about 34 percent (approximately 13 million tons) of the material is polluted. The largest part of the dredging is done directly by the Corps; the remainder is done by private contractors under Corps permits. Often spoils are deposited in open coastal waters less than 100 feet deep.

909. What are *dredge spoils*? *Dredge spoils* refers to the solid materials removed from the bottom of water bodies, generally for the purpose of maintaining channels or improving navigation. Such materials may include sand, silt, clay, rock, and various pollutants that have been deposited from municipal and industrial discharge outlets.

910. Do dredge spoils pollute in ways other than reducing the oxygen demand? Yes. Many water areas where dredging is required to maintain navigation channel entrances to port areas also receive waste discharges from industrial plants producing metal products. Over a period of time, these heavy metals will tend to concentrate in the bottom materials and, when dredged, are mixed again with the ocean (or lake) waters. For example, an analysis of dredge spoils from an area off a Lake Erie port revealed significant concentrations of cadmium, chromium, lead, and nickel.

911. What are the major chemical pollutants? Sugar refining plants dump sodium chloride and sodium sulfide into the water. Foundaries dump chlorides and sulfates. Printing establishments contribute inks and dyes. Electroplating plants add cyanides and hydroxides.

Fertilizers, pesticides, fungicides, herbicides, and other agricultural chemicals are washed into streams in increasing quantities. Detergents, mercury, lead, and copper all find their way into the sea.

912. Why are nitrogen, phosphates, and organic materials in waste water harmful? Algae are simple plants present in most water in a natural state. While growing in the sunlight, they supply oxygen to the water. When they become overabundant in heavily fertilized (i.e., with nitrogen, phosphates, and organic wastes) water, they die and decay; oxygen normally dissolved in the water is used up in the process. Heavy algae growth and decomposition create objectionable odors, cause color and bad taste in drinking water, are unsightly, kill fish and other aquatic life by using needed oxygen, and sometimes create a toxin deadly to animals, including man.

913. Do the oceans have an unlimited capacity to dilute wastes? Because of the vast volume of waters in the oceans, the continuing mixing process and biological decomposition, many kinds of wastes can be dispersed in the open ocean. There are, however, some substances which are concentrated by marine organisms to the point where they can be hazardous to man. Among these are the radionuclides and mercury. There are also materials such as lead which do not dissolve readily in water. Lead from automobile exhaust entering the ocean today will still be present centuries from now.

In contrast to the open ocean, coastal areas are easily polluted to the point where they cannot be used for swimming or harvesting shellfish.

Some chemical pollutants destroy bacteria in restricted waters and prevent self-purification of the water. Among these are the phenols, a constituent of coal tar.

914. What is the definition of the term *solid waste* when referring to ocean dumping? *Solid waste* is probably most commonly called garbage, trash, or refuse. It consists of unwanted and discarded material generated by residences; commercial, agricultural, and industrial establishments; hospitals and other institutions; and municipal operations which receive wastes that include chiefly paper, food scraps, garden wastes, steel, glass, and plastic containers, and other materials.

915. What materials are included in construction and demolition debris which are dumped in the oceans? In ocean dumping and pollution technology this term includes mainly masonry, cinderblock, plaster, tile, but can also include wiring, piping, plastic, tar, roofing materials, glass, vegetation, and excavation dirt. Wooden debris is not usually included.

916. How many ocean dumping sites are there off the U.S. shores? According to figures collected and reported by the Council of Environmental Quality in 1970, there are almost 250 disposal sites off the coasts of the 48 contiguous states. (No information is available for Alaska and Hawaii.) Fifty percent of the disposal sites (122) are located off the Atlantic Coast, 28 percent (68 sites) are off the Pacific Coast, and 22 percent (56 sites) off the Gulf Coast. These sites do not include about 100 artificial reefs constructed from hard rubble (solid waste) which were specifically developed as sport fishing areas.

917. What Federal and/or state agencies have regulatory authority over ocean pollution and dumping? By 1970 it was realized by many in positions of authority that current laws, regulations, agency policing authority and designation of responsibility were inadequate to handle the problem of ocean dumping. Several corrective steps have been taken recently at both state and Federal levels to improve the situation. In 1970 a few states began to require permits for ocean dumping, though state authority is generally limited to the 3-mile territorial sea (except for the west coast of Florida and the Texas coast where it is 9 miles). At Federal levels, the Army Corps of Engineers' authority to regulate dumping is primarily restricted to the territorial sea. The Coast Guard enforces several Federal laws regarding pollution but has no authority to regulate ocean dumping. Likewise the Federal Water Quality Administration (or its new replacement, the Environmental Protection Agency) does not have regulatory authority for ocean dumping. The Atomic Energy Commission does have authority to oversee and monitor disposal of radioactive materials. It is now a well-recognized fact that new legislative authority is needed. More important, probably, is the fact that Federal officials also recognize the international character of ocean dumping; they suggest that cooperative international action is mandatory if the problem is to be corrected on a long-term basis. Unilateral action by the United States can be only a partial solution.

918. Can pollutants be successfully treated before reaching the ocean?   Municipal sewage is relatively simple to treat, but adequate treatment plants will be very expensive. Industrial wastes are more difficult to treat, but the sources of pollution are easily identified. Probably the most difficult form of pollution to control is wastes which come from widespread sources. These include pesticides, herbicides, and fertilizers; chemicals used on icy roads; and lead from automobile exhausts. All of these find their way into the oceans.

919. Do most municipal waste treatment plants treat wastes adequately?   In 1969 less than ⅓ of the nation's cities and towns were reported to have adequate sewer systems and treatment plants. Almost 40 percent have no sewer system at all or discharge waste into natural waters without treatment! The remaining ⅓ of the municipalities have sewers, but inadequate treatment plants. The greatest municipal waste problems are in areas of heavy urban population concentrations, most of which are located along coastal areas of the Northeast, the Great Lakes, and the Western states. The outlook for correcting and staying ahead of requirements is not encouraging, inasmuch as waste loads are predicted to quadruple over the next fifty years!

920. What is the pollutant called *sewage sludge?* *Sewage sludge* is the solid material that remains after municipal waste water has received treatment. It includes human organic wastes and other wastes, both organic and inorganic.

921. Can the oceans absorb large quantities of sewage without harming the ecological balance? There are an estimated 12 billion cubic feet of ocean for each person on earth. Biochemical processes can purify the waste if the concentration is no greater than 1 part sewage to 200 parts seawater. The problem is that man has concentrated his cities and industries along the coasts and it is the coastal zones that are being heavily polluted. Sewage sludge contains high concentrations of toxic metals. In addition to municipal sewage, there are an estimated 1.3 million boats with toilets in the United States, according to the Department of the Interior.

922. In what coastal areas is sewage sludge disposal in the ocean a serious problem? According to the statistics for the year 1968, only about 4.5 million tons of sewage sludge were dumped at sea. Of this, about 4 million tons were dumped by New York City in coastal waters! Sludge is discharged by Los Angeles by a pipeline for which no records are available. Dumping operations like New York City's are not practiced elsewhere along U.S. coasts.

923. Why are coliform bacteria considered indicators of polluted water? Coliforms are a group of bacteria common to the intestinal tract of warm-blooded animals. Although some forms live and reproduce outside the intestinal tract, their presence in water is a good indication that fecal material and possibly disease germs may also be present. The higher the coliform count, the greater the danger in using untreated water.

924. How extensive is oil pollution from ships? About 1,000 million metric tons of oil are transported by sea every year. An estimated 0.1 percent, or about one million tons per year, is spilled or leaked into the sea. Oil tankers that wash their tanks with salt water and dump the oil residue in the ocean are among the polluters, but passenger ships and freighters also pump out seawater ballast from their fuel tanks before entering port to refuel.

925. What are other sources of oil pollution?   Manufacturing plants, refineries, and oil terminals contribute oil to the oceans. The Massachusetts Division of Natural Resources states that several tons of oil can be expected to be spilled in Boston Harbor every three weeks.

Seepage from offshore drilling rigs adds to pollution levels. One well in California's Santa Barbara Channel leaked 236,000 gallons over a ten-day period.

926. What is the source of most of the oil spills?   It is estimated that there are about 10,000 incidents in which oil (or other hazardous materials) are spilled into navigable waters of the United States each year. In a tabulation of the source of oil spills of 100 barrels or more occurring in 1969, the Federal Water Quality Administration (now the Environmental Protection Agency) reported that one half of the spills were from vessels and one third of the incidents involve pipelines, oil terminals, and bulk storage facilities. Vessels, especially tanker casualties, are widely publicized pollution sources because millions of gallons of oil are spilled into the water at one time and place, and the effect can be catastrophic. Blowouts and damage or destruction at offshore drilling rigs are also potential sources of large spills. One little-known fact is that gasoline service stations dispose of more than 350 million gallons of used oil each year into water drainage systems!

927. Does oil pollution occur naturally?   Oil from natural seeps has been described in government reports since 1902, but no scientific investigations were conducted until recently. Texas A&M University's Oceanography Department is now studying natural oil seeps in the Gulf of Mexico. The research is supported by the Sea Grant Program of the National Oceanic and Atmospheric Administration and grants from six oil companies. Field surveys include hundreds of drift cards, released in seep areas, to analyze the path and time for water movement from the sites.

928. What are the chances for the entire ocean to be polluted by oil spills? Reliable sources estimate that approximately 60 percent of the world's oil is transported on the ocean and that only about $\frac{1}{10}$ of 1 percent is spilled or leaks from offshore drilling rigs. However, marine biologists aboard the Woods Hole Oceanographic Institution research vessel *Chain,* making collections with their neuston nets in the Sargasso Sea during November and December 1968 had cause for concern. In towing these nets for 30 minutes to an hour along the sea surface (covering approximately 2,000 to 3,500 meters) they collected a cupful or more of oil-tar lumps. The size of the lumps ranged from droplets to about 5 centimeters. It is also understood that similar oil-tar collections have been made in the South Atlantic Ocean and the Caribbean. It is uncertain at this time whether this oil comes from undersea wells, is from natural seepage, or from ship bilges. Though no definite conclusions can be drawn from the few random collections similar to those taken by the *Chain,* it is suspected that other oil-tar concentrations exist but have been unreported for many ocean areas. The whole ocean certainly cannot be construed as being polluted from this evidence, but these substances in the Sargasso Sea can be damaging and perhaps affect the ecology since the sargassum weed floating at the surface would tend to collect the oil-tar blobs much as the net did. This "weed" serves as the habitat for some 16 species of marine creatures.

929. How costly is oil pollution? According to a report by Arthur D. Little, Inc., oil spills at sea may cost as much as $10 million per year. The cost of cleaning up oil spilled when the tanker *Torrey Canyon* went aground in 1967 was between $8 and $9 million. Many of the beaches and coasts of England were still contaminated several years later.

930. What happens to oil spilled at sea? Crude oil spilled at sea will disappear within a few weeks without any help from man. The more volatile portion evaporates and the remainder is decomposed by bacteria into water and carbon dioxide. The time required for decomposition depends on the type of crude oil and temperature of the water. The Coast Guard is conducting studies to determine how long oil spills will persist in the Arctic.

931. How does oil affect marine birds? Along the coastlines of
the Northern Hemisphere, perhaps a million waterfowl of various
species die a lingering death every year because of oil pollution. Oil
mats the feathers and allows cold water to reach the skin; as a result,
the birds finally freeze to death. Oil may rub off from the feathers of
nesting birds onto their eggs, clogging the pores of their eggs and suf-
focating the embryo.

A coating of oil may make flight impossible and the birds become
easy prey to predators.

932. How do detergents affect marine life? Marine life along
the coasts of Britain was damaged more by the 2 million gallons of
detergents used to clean up oil from the *Torrey Canyon* spill than by
the oil itself. Less toxic detergents are now used to remove oil from
the sea surface and disperse it into the sea. Some scientists believe
that detergents will cause serious damage to subsurface organisms
which would not be affected if the oil were left on the surface; they
suggest that no attempt be made to clean up oil spills.

933. Do oil spills affect plankton and fish? Oil spilled on the
sea quickly loses its more volatile components, which might be toxic
to marine life. It has no measurable effect on either plankton or fish,
although both fish and shellfish may be tainted by oil and become unappetizing to man.

**934. How many types of hazardous materials constitute a
threat to water environmental quality?** Results of an open sym-
posium sponsored by the Coast Guard in September 1970 turned up
many chemicals, biologicals, and pharmaceuticals as potential pollutants. A total of about 700 hazardous materials of this type were
listed, of which 300 were listed for "serious consideration" by the
sponsors. Records show spills of 200 of these different materials
each year but during the meetings one estimate of the number of occurrences was 8,000 per day! Two examples of hazardous chemicals
which are widely transported and can create severe contamination
problems are anhydrous ammonia and carbon tetrachloride.

**935. What are some of the toxic industrial wastes dumped at
sea?** Industrial wastes dumped at sea come from a great variety of

industries and production processes. Some of the more toxic substances dumped at sea include cyanides, heavy metals, mercaptides, chlorinated hydrocarbons, and arsenic and mercuric compounds. The most toxic are often put in 55 gallon drums and taken some 300 or more miles offshore for disposal. Some containers are weighted and sunk; more often the drums are sunk by rupturing them while at the surface by axes or rifle fire.

936. How widely are pesticides distributed in the oceans?

Pesticides used in great quantities in agriculture are washed into streams and rivers which carry the materials into the sea. Currents flow constantly through all the oceans, both horizontally and vertically. Persistent pesticides, such as DDT, have been found in all parts of the oceans. Some scientists believe that two thirds of the 1.5 million tons of DDT produced by man may still be in the oceans. Pesticides used on the African Continent have been found in the Bay of Bengal and the Caribbean Sea.

Pesticides are also carried through the atmosphere, and it may be that a major share of the pesticide reaches the offshore parts of the oceans through the air.

937. How do toxic chemicals, pesticides, and other pollutants that are mixed in seawater to only 1 part in a billion or less become concentrated to dangerous levels in fish and other edible seafoods?

Pollutants in very, very dilute concentrations in seawater tend to be reconcentrated in the natural oceanic food chain. The direct effect of 1 pound of DDT in 10 million pounds of seawater on most organisms would be negligible. Many marine organisms, however, filter out such chemicals as DDT or mercury from the water and collect them in certain body tissues where concentrations become much higher than the water in which they live. In many cases, increased concentration of the chemical in the body tissues may not noticeably affect the activities of the organisms. The smaller organisms are then eaten by larger marine animals and so it continues in succession up the food chain with each organism in turn increasing the concentration of certain chemical pollutants in its body until a toxic level is reached. A good example of food chain reconcentration of a chemical occurred in 1957 in Clear Lake, California. In the lake, the concentration of the pesticide DDD (similar to DDT) was measured at 0.02 parts per million. Plant and animal organisms in the lake were reported to have stored residues of the pesticide at 5 parts per million (250 times the concentration in the lake). Since fish of the lake consumed great quantities of these organisms, DDD accumulated in the fish was in excess of 2,000 parts per million, more than 100,000 times the original lake concentration. By this time, accumulations were at toxic levels for birds which ate the fish, and many died.

938. Is it possible to make fish caught in polluted waters edible?   Maximum allowable limits are established by the Federal Government for various chemical substances in fish sold commercially for human consumption. It is illegal to sell for food, fish or fish products which contain pollutants concentrated in the fish in amounts greater than the Federal Food and Drug Administration permits. It is not generally recommended that one eat fish or shellfish taken from waters which are identified and posted by health authorities as polluted. However, due to the nature of fish physiology, procedures can be used during the cleaning process to reduce to some extent concentrations of some chemical contaminants found in fish. For example, 5 parts per million (ppm) of DDT is the maximum allowable in fish sold commercially. Some coho salmon in the Great Lakes have been found to contain residues of DDT up to 19 ppm. It has been found that much of the DDT can be eliminated during the cleaning process if the fat, where most of the DDT is concentrated, is trimmed off. Generally, eliminating the fatty parts running the full length of the belly, the area just under the dorsal fin and to the rear of each gill, will substantially reduce the overall concentration ratio.

939. How do pesticides affect phytoplankton?   Scientists of the Woods Hole Oceanographic Institution have found that as little as 10 parts per billion of DDT, endrin, or dieldrin can slow the growth of some marine plants. Any decrease in plant growth threatens the supply of oxygen in the ocean and atmosphere.

940. Why are oysters so susceptible to pesticides?   Oysters and other shellfish have only limited mobility and cannot easily escape polluted areas. They circulate water through their bodies as a means of obtaining food. Toxaphene and dieldrin have been found to be especially irritating to oysters, but there is some evidence that the damage is not permanent. Eggs and larvae are especially susceptible.

941. How does DDT affect sea birds? DDT enters the food chain in marsh waters from farm runoff and becomes concentrated in fish-eating birds. It has interfered with the reproductive cycle of the osprey and peregrine falcon and has been found in large quantities in Antarctic skua, Atlantic and Pacific shearwaters, and Bermuda petrels.

942. What radioactive wastes might be included as possible marine pollutants? Radioactive wastes dumped or discharged into the ocean which could possibly pollute marine life and water, consist of liquid and solid wastes resulting from processing of irradiated fuel elements, nuclear reactor operations, medical use of radioactive isotopes, and research activities. Equipment and containment vessels which become radioactive by induction can also be another source of pollution.

943. Is the disposal of radioactive wastes in the ocean by the United States increasing? No. Since 1962 the disposal of radioactive wastes at sea has been dramatically reduced. A moratorium on granting new licenses to authorize disposal of radioactive waste in the ocean was placed in effect by the AEC in 1960; in 1970 only four activities had licenses permitting disposal at sea. The record shows a reduction in the number of containers with radioactive wastes disposed at sea from 4,087 and 6,120 for the years 1961 and 1962, respectively, to 26 and 2 for 1969 and 1970. Today, land disposal of radioactive wastes is the preferred procedure.

944. Is the water discharged from nuclear power plants safe? The water is safe enough to drink, but some of the plants and animals that live in it may concentrate radioisotopes to several thousand times that in the surrounding water. Some scientists are of the opinion that any discharge of radioactive wastes is potentially harmful, especially if it reaches the sea at times when fish eggs are developing.

945. How does the radioactivity of sea and land compare?
The natural radioactivity of the sea is less than $\frac{1}{50}$ of that present in either sedimentary or granitic rocks.

946. Is it safe to dump radioactive wastes in the deep ocean?   The answer to this question is not yet known with any certainty. Vertical and horizontal currents are believed to be sluggish in water deeper than a mile; it might take centuries for radioactivity to reach the surface. On the other hand, recent discoveries of deep countercurrents have caused some scientists to doubt the sluggishness at great depths. The Russians have vigorously opposed any suggestion of marine disposal of high-level wastes.

947. How long has the United States practiced disposal of munitions at sea?   The practice of getting rid of unserviceable and obsolete shells, mines, solid rocket fuels, and chemical warfare agents has been standard procedure for many years. It was in 1963 that the Department of Defense began to use outdated World War II liberty ships as the munition carriers, then simply scuttling the ships after they are towed to the designated disposal area. Over the past seven years, 19 old ships have carried over 53,000 tons of war materials to the ocean bottom. Of this, approximately 20,000 tons were explosives and munitions.

948. Does the ocean contribute to air pollution?   Until recently, scientists believed that nearly all atmospheric carbon monoxide was produced by man-made processes. It has recently been demonstrated that as much as 10 percent of the carbon monoxide in the atmosphere comes from the sea surface. There is a marked increase in the dissolved carbon monoxide at the sea surface in late afternoon, suggesting a biological source. Scientists of the U.S. Naval Research Laboratory have suggested that the source of carbon monoxide in the sea is either siphonophores in the scattering layer or photochemical oxidation of organic material in the upper layers of the ocean.

949. Why is carbon dioxide increasing in the oceans?   Since the beginning of the industrial revolution, burning of fossil fuels has produced carbon dioxide, sulfates, and nitrates, much of which enters the sea. The United States is responsible for more than one third of the world's total. Solid smoke particles accumulating in ocean bottom sediments may alter the chemical composition of the ocean.

950. How do lead and mercury get into the ocean? Both heavy metals enter the ocean in part through man-made discharge. An estimated 10,000 tons of lead are introduced annually. Since lead was introduced into gasoline less than fifty years ago, lead concentrations in the Pacific surface waters have increased by a factor of 10.

Some 4,000 to 5,000 tons of mercury enter the oceans every year by natural erosion; man introduces an equivalent amount. Mercury accumulates in plants and wide-ranging fish such as swordfish and tuna.

951. What concentrations of heavy metals contamination are detrimental to marine life? A comparison of natural seawater concentrations of several heavy metals, concentrations toxic to marine life, and concentrations in dredge spoils follows:

Metal	Concentration toxic to marine life	Concentration in natural seawater	Concentration in dredge spoils
	– In parts per million –		
Cadmium	.01 to 10.0	.08	130
Chromium	1.0	.00005	150
Lead	.1	.00003	310
Nickel	.1	.0054	610

952. What are the effects of copper on organisms? Large quantities of copper are discharged through the drains of all industrialized countries every year, ending in the ocean. Some animals, such as oysters, tend to concentrate copper. A relatively large quantity of copper (0.13 part per million) will not kill oysters but can give them an unappetizing green color. Scientists are not in complete agreement on the possible harmful effects of copper on marine organisms.

953. Is there any danger from poison gases dumped in the oceans? More than 30,000 bombs and canisters containing poison gases were dumped in more than 90 meters (295 feet) between Sweden and the Danish island of Bornholm after World War II. On August 10, 1969, the London *Sunday Times* reported that mustard gas, leaking from the rusting canisters, had injured six fishermen seriously and brought panic to holiday resorts in southern Sweden and Bornholm. Thousands of tons of fish suspected of contamination were boycotted. Danish authorities expressed the belief that the containers were shifted by tides, currents, or trawler's gear into shallow water off the Baltic resort areas.

954. Is the rate of decomposition known for the nerve gas (rockets) dumped into the ocean in 1970? Edgewood Arsenal (Maryland) scientists have recently completed tests on the decomposition rate of nerve gas in seawater. The poison is often referred to as GB; the chemical name is *isopropyl methylphosphonofluoridate*. It reacts in water to form a harmless substance, *isopropyl methyl phosphionic acid*. Data from an extensive series of experimental tests performed by Joseph Epstein at the Arsenal show the half-life of the poisonous agent to be about 16 hours. This means that one half of the chemical is reduced to the harmless acid when released into cold seawater similar to the deep ocean environment where it was dumped. Chemists found that the calcium and magnesium ions in seawater tend to accelerate the decomposition of the poison.

955. How does erosion contribute to pollution? Fine sediments high in organic content have been eroded from uplands and deposited in the coastal zone, ruining many east coast swimming areas. Spoil from dredging can also contribute to the deterioration of swimming areas. Marine life is suffocated by fine silts washed into shallow nearshore areas.

In southern California there is an opposite condition. Sand supplied by rivers is transported along the beach by wave-generated currents and lost to the deep sea canyons. Runoff control has reduced erosion and decreased the supply of sand; as a result, beaches are becoming seriously depleted.

956. What effect does pollution have on shellfish culture? More than 1.2 million acres of shellfish grounds have already been declared health hazards by the Public Health Service. This represents 8 percent of the shellfish grounds of the United States. Most of the harbor and estuary areas suitable for oyster farming are already polluted beyond the point where they could be safely used.

957. Can brine from desalination plants be dumped into the ocean without ecological damage?   The problem resulting from dumping concentrated brine into the oceans is a serious one because many organisms are sensitive to changes in salinity. Scientists of the University of Puerto Rico have studied the effects of dumping a concentrated brine from a salt works in southwest Puerto Rico. The brine destroyed all bottom organisms and produced hydrogen sulfide and patches of acid water with a pH as low as 3.5.

958. Why is dissolved oxygen so important in water?   Oxygen is needed to "burn up" or decompose organic pollutants in an odor-free manner. Its presence in water is also vital to fish and other forms of desirable aquatic life.

959. What does *BOD* mean?   *BOD* is an abbreviation for *b*iological *o*xygen *d*emand and is a term used to express the amount of dissolved oxygen needed to "burn up" or decompose the organic material in a particular volume of water.

960. What is thermal pollution?   Thermal pollution is the addition of substantial quantities of heat to natural bodies of water, usually by cycling huge volumes of water through large power plants for cooling purposes and pumping the heated water back into the stream, river, lake, or ocean from which it was taken. If the body of water is a stream, small river, or lake, the normal ambient temperature may be increased by several degrees or more at the point of heated water discharge and, depending on volume and rates of use, may extend over many square miles.

961. How do nuclear power plants increase thermal pollution?   Safety requirements for lower steam temperatures and pressures make nuclear power plants less efficient than coal-fired plants. As a result, they heat up the water about 40 percent more than coal-fired plants which transfer some of their heat into the atmosphere.

962. What is the outlook for the increase in thermal pollution in U.S. waters in the 1970s and beyond? According to statistics obtained by the Council on Environmental Quality, the electric power industry (the source of most thermal pollution in the United States) is growing at a rate of 7.2 percent annually. The Council predicts that this trend will continue; this means that the electrical output is expected to double, approximately, over the next ten years. The principal use of water in generating electricity is for cooling the condensers in the steam-electric generating plants, most of which are fossil-fueled. Usually the temperatures of cooling water are increased between 10° and 20° F. Large nuclear steam plants, however, require about 50 percent more condenser cooling water for a given temperature increase than do fossil-fueled steam electric plants of the same size. Increased use of nuclear energy power plants indicates a possible additional drain on existing natural waters for cooling in the coming decade. By 1980 it is estimated that the electric power industry will require the equivalent of one fifth of the total fresh-water runoff of the United States for their cooling operations.

963. How does thermal pollution disrupt the ecology? Increased temperature decreases the oxygen-carrying capacity of water and may change the ecology. Warmer waters might cause a fish hatch to occur so early in the spring that the fishes' natural food organisms would not be available. Aquatic plants, the primary food, depend on day length as well as temperature.

The sensitivity of all aquatic life to toxic materials is heightened at increased temperatures. Organisms adapted to water which seldom becomes warmer than 70° F may be killed by 90° F water. Even at 75° or 80° F they may be unable to compete with other species better adapted to higher temperature. Trout eggs will not hatch in water that is too warm and salmon may not spawn.

964. What beneficial results may come from thermal pollution? Warmer water may improve an area's potential for recreation, and it may also stimulate aquaculture programs. Under certain circumstances the warmer water may stimulate the growth of crabs, shrimp, and fish.

Off southern California, waste heat has been maintaining a year-round population of subtropical fishes that are usually present only in summer.

965. How could thermal pollution from nuclear plants be avoided?

Some nuclear power plants are combined with desalination plants, so that waste heat can be used in desalting water.

Heated cooling water from nuclear power plants could be discharged at depths, rather than at the surface. The heated water would then rise to the surface, bringing with it nutrient-rich deep water.

966. What would it cost to clean up water pollution in the United States?

The cost over the five-year period 1970–75 of building needed municipal waste treatment plants that would meet present water quality standards would be $10 billion. Operational costs over the same five years would average $560 million per year and will total almost $3 billion for that period. Estimates for correcting problems of sewer overflows, made by the American Public Works Association, are placed at $15 to $48 billion, depending on the method chosen.

Excluding thermal pollution, the price for correction of industrial contamination of natural waters to meet established water quality standards by 1975 is estimated at $3.3 billion. Another $2 billion would be needed in the next five years to provide recirculation water systems to prevent thermal pollution. The total estimated direct costs for correcting existing pollution problems so that water quality standards are met is substantial; it totals at least $33 billion and could be as much as $65 billion. There are in addition many indirect costs that involve modification of existing buildings and construction, and correction of agricultural and water management practices. Experts are reluctant to estimate the extent of this work and its cost.

967. Who is responsible for purity or quality of drinking water in the United States?

The standards for drinking water quality were established more than fifty years ago—in 1914—when Congress passed a law requiring interstate quarantine regulations to be established by the Surgeon General who in turn established the U.S. Public Health Service Drinking Water Standards. These are actually performance specifications which in effect state goals. There are only about 650 places in the United States where interstate carriers take on drinking water; these are the only locations subject to Federal inspection, laws, and regulation. According to James H. McDermott, director of the Bureau of Water Hygiene, the states have responsibility for the well-being of all the people in each state and authority relating to the quality of drinking water in towns or urban dwellings. He indicates that no one has challenged the Federal standards or the use of them. Since the original enactment in 1914 there have been revisions in 1925, 1942, 1946, and 1962. Each of the first three placed acceptable drinking water limits on inorganic chemicals; the 1962 revision placed limits on organic chemicals. Since 1967, efforts have been made to amend the standard again, this time to include pesticides.

968. What are some possible water supply sources if shortages occur in the future?   Three possibilities are frequently mentioned by water experts as practical solutions to future water shortages. The first, desalination of seawater is becoming more extensively used each year to supply large urban areas with needed fresh water. As the engineering of desalination processing plants improves, the cost per gallon for production should be lowered and these systems should find increased use to supplement the water systems of coastal cities. A second solution, complete recycling of existing water, is becoming more and more a reality every day in some areas. However, a basic problem is the "chance for error"; a true "fail-safe" system in the cycle is an absolute requirement. The third solution is a dual supply concept. This is not a new concept as it is used on many ships. It provides for two supply (piping) systems. One source would be used for drinking, cooking, bathing, and other bodily contact needs which account for only 20 percent of our water use. The second system would take care of other uses such as water for the lawn, car washing, etc.

XV. MYTHS AND LEGENDS

969. Do sunken ships drift forever a mile below the surface? It was once widely believed that the great pressure at depths made the water so dense that objects sinking to a certain depth would be held there forever. Only a hundred years ago, sailors of the British oceanographic survey vessel *Challenger* asked Sir John Murray, the expedition leader, whether their dead shipmate with a shot attached to his feet would sink to the bottom or float at some level. Murray assured them that for all practical purposes anything that will sink to the bottom of a tumbler will sink to the bottom of the deepest ocean. Water, even under tremendous pressure, is almost imcompressible; wood, on the other hand, will be compressed to half its volume at a depth of 3,000 feet and will sink.

970. Can ships become trapped in the Sargasso Sea? Legends of ships becoming entangled in masses of sargassum weed while their crews became mad or died of thirst were believed by many sailors in the days of sailing ships. There is no basis of fact to these legends; the floating weeds could not trap even a small sailboat.

971. Are there areas of "dead water" that can cause a steamer to stand still? This can happen in areas where there is a thin layer of low density water over more dense water. This condition exists in the Arctic and areas of river outflow. If the water layer is no deeper than the ship's hull, waves are generated between the boundaries of the two layers. The energy of the ship's screws goes into generating and maintaining these internal waves instead of thrusting the ship ahead. Increasing the number of turns will usually restore forward movement.

972. What is the legend of Atlantis? The story of Atlantis is one of the most persistent of all legends. More than 5,000 scholarly works are in existence about the lost Atlantean civilization. The oldest existing account is that of Plato and this is the basis for most sub-

sequent speculations. The story of Atlantis was also told by Homer
in the seventh century B.C. and Herodotus in the fourth.

Plato described a great civilization far ahead of its time, located
west of the Pillars of Hercules. The Atlanteans built temples, ships,
and canals. They lived by agriculture and commerce. In their pursuit
of trade, they reached all the countries around them. By 9600 B.C.
they had conquered all the known world except Greece, which was
saved when Atlantis was engulfed by the sea, overnight, disappearing
without a trace.

973. Is there evidence to support the Atlantis legend?   In recent years there has been speculation that the location of Atlantis
was 70 miles north of Crete where Santorini erupted between 1500
and 1450 B.C. with a violence at least ten times that of Krakatoa.
The present island of Thera is the largest fragment remaining after
the eruption and subsequent seismic sea waves.

It has been suggested that a translation error placed the destruction of Atlantis at 9,000 years before Plato received the story from
Solon (590 B.C.) instead of 900 years. If this assumption is correct,
the destruction of Atlantis would have been in 1490 B.C., corresponding with the eruption of Santorini. The numbers of soldiers and
sailors and the dimensions of the island given by Plato would also be
more plausible if divided by 10.

The area around Thera has been surveyed by oceanographic research ships of the Woods Hole Oceanographic Institution and Lamont-Doherty Geological Observatory. Excavations of the island are
now under way.

974. What sea monster is reputed to swallow ships?   In sailors' lore the kraken was a sea monster that could swallow vessels
whole. Many stories have been told of sea monsters pulling men from
ships and dragging them to their death. The creatures responsible for
these tales may have been giant squids which can have tentacles more
than 40 feet long.

Oarfish 40 to 50 feet long have been observed by scientists. These
may have been reported as sea serpents by sailors of old.

975. When was the first sea monster reported?   The first written account of a sea monster was recorded by Aristotle in the fourth

century B.C. Since then, many sea monster sightings have been reported. In the sixteenth and seventeenth centuries, two Scandinavians, both clergymen, recorded observations of huge sea creatures told to them by fishermen and mariners; they were Olaus Magnus, Archbishop of Uppsala, and Erik Pontoppidan, Bishop of Bergen. Another minister, Hans Egede, who was an eighteenth-century missionary to Greenland, reported sighting a sea monster while en route to Greenland; he recorded the event in considerable detail.

976. When was the Loch Ness monster first seen? On April 14, 1933, John McKay and his wife, while driving along the north shore of Loch Ness, saw this huge, strange, and now-famous creature. They reported seeing an animal with a huge dark body and a long snakelike neck. According to their statement, it surfaced, swam along for a few seconds, and disappeared, making considerable foam and splash. When reported in the press, it received world-wide attention; in the years since, others persons have also reported seeing the Loch Ness monster. It has now become the best-known and most frequently sighted sea monster. Actually, since Loch Ness is a lake, the term *sea monster* might not be entirely appropriate.

977. Have carcasses of sea monsters ever been found? There are many reports of dead sea monsters stranded on beaches; some have even been photographed. Most of those examined by scientists have been basking sharks, which sometimes reach a length of 60 feet. Dead basking sharks decay quickly on the beach, leaving a residue of backbone, muscle, and cartilage which have often been reported as sea monsters.

978. Are there any fossil records of sea serpents? The only creature known to science that qualifies as a sea serpent is the zeuglodont, a primitive whale. Its fossil remains have been found in Alabama and Florida. It was over 60 feet long, had a snakelike body and a tail similar to that of a whale. The creature lived during the Eocene epoch.

979. Do any reputable scientists believe in sea monsters? The late Anton Bruun was a highly respected marine biologist who believed that such creatures do exist. During oceanographic expeditions on the Danish vessel *Galathea* he searched for them unsuccessfully. He did not, however, neglect his primary scientific investigations.

J. L. B. Smith also expressed the belief that sea serpents exist in the deep sea. He is the famous ichthyologist who described the coelacanth discovered in 1938. This fish had been thought to be extinct for the past 50 to 70 million years.

980. Is there any scientific evidence to support a sea serpent of monster size?   A startling incident occurred in 1930, when the distinguished Danish marine biologist Dr. Anton Bruun was working off South Africa. One of his trawls brought up a leptocephalus (eel larva) which was 6 feet long. The normal size of eel larvae at this particular stage in the life cycle is usually no more than 4 inches. If this 6-foot specimen had continued to grow to maturity, assuming development at a proportional rate of the normal eel larva to the mature eel, it would have developed into an adult specimen approximately 70 feet long. This specimen is considered by some individuals to be evidence of the existence of a sea serpent.

981. What are the horse latitudes?   This is a picturesque name, dating from the days of sailing vessels, for the calm areas centered 30° to 35° north and south of the equator. Winds are calm or very light and weather is hot and dry. The origin of the name is obscure. One story is that ships became becalmed and horses died of thirst and were thrown overboard; another version says that horses died because of the unfavorable climate.

982. Are undertows dangerous to swimmers?   The old myth about the dangers of undertows has been dispelled by scientific studies at Scripps Institution of Oceanography. There is some slight outward movement at the bottom, but the dangerous currents which cause drownings move seaward at the surface.

983. Is every seventh wave higher?   This is an old myth which has been kept alive in recent years by surfers. The larger waves which arrive on the beach periodically, result when two crests from different directions become superimposed and form a higher wave.

Although there may be regularity for a time, the pattern changes constantly.

984. Did surfing originate in Hawaii?   According to available records, the first observation of surfing was made by Captain Cook in 1778. He writes in the ship log, "The boldness and address with which we saw them perform these difficult and dangerous maneuvers was altogether astonishing and scarcely to be credited." Later, after Cook's death, his successor, Captain King, went to great lengths to describe the sport which he found in Hawaii. The origin of surfing, however, goes back many centuries. Indications are, from the many references to surfing in Hawaiian chants, that it was practiced throughout the Pacific. The highest development of skill in surfing seems to have been attained in Tahiti and Hawaii. On other Pacific islands, from New Zealand to Easter Island and New Guinea, surfing on boards was left to children. Based on an analysis of the migration of Polynesians, a study of their chants and of surfing as performed by the native populations when first observed, it appears that surfing originated in Tahiti and came to Hawaii when Tahitians found their way there, using stick charts, more than 1,000 years ago.

985. Did oceanography start as a rich-man's sport?   There is some truth to this belief. Before the days of government funding, only men with comparatively large personal fortunes could afford the expense of shipboard science. These men were not playboys, however; they were dedicated scientists.

Alexander Agassiz spent more than $1.5 million on the study of oceanography and marine biology. Prince Albert I of Monaco also spent a fortune and served as captain and chief scientist of four yachts which he sailed in the Mediterranean and North Atlantic.

Columbus Iselin, who died in 1971, was able to begin his studies of the ocean because he could afford the sport, but he became a respected scientist who served for many years as director of the Woods Hole Oceanographic Institution.

986. Are oysters good to eat only in months with an "R"?   Oysters can be eaten any month of the year, but they are fatter and tastier in the spring. From the viewpoint of quantity and quality, late spring would be a much better harvest time than autumn.

987. Does blood in the water attract sharks? Tests conducted by scientists at the University of Miami have proved that sounds of struggling animals attract sharks long before the odor of blood could reach them. Sharks can detect such sounds at distances of 200 yards.

988. Do deep-sea fish explode when brought to the surface? Fishes from deepest parts of the ocean never explode when brought to the surface because they have no swim bladders or air spaces of any kind. The change in pressure does not harm them, but the temperature change can be fatal.

It is the mid-water and shallow-water fishes that sometimes explode when brought to the surface. As pressure decreases, their swim bladders swell up and protrude from their mouths.

989. How did Christianity contribute to the naming of a fish? When Christianity spread to the Netherlands and Germany, converts followed the religious custom of eating fish on Friday and holidays. These people found the *butt,* a large flat fish native to the surrounding waters, particularly suitable for eating on those days. As it became the practice to eat this fish on religious days, it became known as *holy butt,* which was subsequently modified to *holibut* and *halibut.*

990. How did sardines get their name? The name *sardines* was applied to the small fish of the herring family which appeared in thick schools off Sardinia. That island was occupied and invaded by Phoenicians, Greeks, Romans, and many other ancient warriors who also knew the fish as *sardene,* the Greek name for the island. The earliest reference in the English language (1430) provides the spelling *sardeynez.*

991. Who named the menhaden? Menhaden was the fish that American Indians used as fertilizer. The Algonquin tribe dropped a fish in each hole with the corn seed they planted. This fish they called *munnohquohteau,* which means "he enriches the land." The first written reference is found in a 1643 volume, where *munnawhatteaug* is the spelling used. A subsequent spelling was *manhadden*.

992. Do porpoises rescue drowning people? There are many stories of porpoises pushing men to shore to save them from drowning. Marine biologists who have studied porpoise behavior point out that porpoises like to push objects around and there is no evidence that they are deliberately being helpful. On the other hand, it is well known that porpoises do seem to like people and hate sharks.

993. Do porpoises eat their own weight in fish every day? This claim is sometimes made, but it has never been substantiated. Porpoises at the Miami Seaquarium are fed 5 times a day and they eat no more than 25 pounds of fish per day.

994. Could a whale have swallowed Jonah? Many pictures of this event show the huge blue whale, the largest of all creatures, as the "fish" that swallowed Jonah. Actually, the gullet of the blue whale is only large enough to swallow small fish; the blue whale feeds on tiny shrimplike krill. The sperm whale (cachalot) appears a more likely possibility. Whalers have watched these animals vomit up huge amounts of recently eaten food when harpooned. There is one record of a giant squid 10 feet long and weighing approximately 500 pounds being discharged. Both the stomach and throat of the cachalot could easily accommodate a grown man.

995. Do whales spout water? It is now believed that a whale blows air, not water. As the air is expelled under pressure it expands, causing water vapor to condense. It is this water vapor which makes the spout visible.

996. Is it possible that a white whale such as Moby Dick can really exist? Albino specimens have been recorded among many animals, so there is every reason to expect such specimens in marine mammals such as whales and porpoises. This expectation was verified recently when a baby white killer whale was captured off the coast of Canada. It measured 12 feet in length and weighed 1,500 pounds. It now lives at Sealand of the Pacific in Victoria, B.C., where it is a companion and mate for another killer whale. Naturally it was named Moby Dick; the other whale is named Haida.

997. Can marine worms cure cancer?   An ancient Polynesian legend attributes cancer cures to a broth made from the marine anne-lid worm *Lanice;* the worm is known to the Polynesians as kaunaoa. Recently, extracts from these worms have been tested at the Univer-sity of Hawaii medical school. It appears to protect mice from cancer when they are inoculated with cells from cancerous mice.

998. Is the smoke screen of octopus and squid used to hide from pursuers?   The pigment released does not make a large enough screen to hide the octopus or squid. Apparently it is used as a decoy to mislead the predator and deaden its sense of smell.

999. How did the mermaid legend originate?   Probably from the fact that females of some marine mammals such as the dugong and manatee hold their young with a flipper while they feed at the breast. These mammals belong to the Order Sirenia. In days of sail-ing ships they were called *sirens;* now they are commonly called *sea cows.*

1000. Is there enough gold in the sea to make every person on earth a millionaire?   Yes, there is enough gold to give everyone a million dollars' worth by today's prices. But if a cheap industrial method of extracting gold from the sea is ever developed, gold would lose much of its value.

1001. What Biblical account probably refers to the red tide?   The red tide and its effects on fish have been known since Biblical times. Dr. Harris B. Stewart, Jr. director of the Atlantic Oceanographic and Meteorological Laboratories of the National Oceanic and Atmospheric Administration, says that probably this particular phenomenon occurred in the lower Nile and is recorded in the Bible in the seventh chapter of Exodus: ". . . and all the waters that were in the river were turned to blood, and the fish that were in the river died; and the river stank, and the Egyptians could not drink the waters of the river."

BIBLIOGRAPHY

General Works

Ericson, David B., and Wollin, Goesta. *The Ever-Changing Sea*. New York: Alfred A. Knopf, 1967.

Gordon, Bernard L., ed. *Man and the Sea; Classic Accounts of Marine Explorations*. Garden City, N.Y.: The Natural History Press. 1970.

Menard, Henry W. *Anatomy of an Expedition*. New York: McGraw-Hill Book Co., Inc., 1969.

Miller, Robert C. *The Sea*. New York: Random House, 1966.

Stephens, William M. *Science Beneath the Sea*. New York: G. P. Putnam's Sons, 1966.

Stewart, Harris B., Jr. *Deep Challenge*. Princeton, N.J.: D. Van Nostrand Co., Inc., 1966.

Marine Geology

Ericson, David B., and Wollin, Goesta. *The Deep and the Past*. New York: Alfred A. Knopf, 1964.

Shepard, Francis. *The Earth Beneath the Sea*. Baltimore: The Johns Hopkins Press, 1959.

Turekian, Karl H. *Oceans*. Englewood Cliffs, N.J.: Prentice-Hall, Inc., 1968.

Waves—Tides—Currents

Bascom, Willard. *Waves and Beaches*. Garden City, N.Y.: Anchor Science Study Series, Doubleday & Company, Inc., 1964.

Bigelow, Henry B., and Edmondson, W. T. *Wind Waves at Sea, Breakers and Surf*. Washington: U.S. Navy Hydrographic Office, 1947.

Clemons, Elizabeth. *Waves, Tides, and Currents*. New York: Alfred A. Knopf, 1967.

Sea Ice

Dyson, James L. *The World of Ice*. New York: Alfred A. Knopf, 1962.

Weeks, Tim, and Maher, Ramona. *Ice Island*. New York: The John Day Company, 1965.

Air-Sea Interaction

Blanchard, Duncan C. *From Raindrops to Volcanoes*. Garden City, N.Y.: Doubleday & Company, Inc., 1967.
Dunn, G. E., and Miller, B. I. *Atlantic Hurricanes*. Baton Rouge, La.: Louisiana State University Press, 1960.
Williams, Jerome; Higginson, John J.; and Rohrbough, John D. *Sea and Air*. Annapolis, Md.: United States Naval Institute, 1968.

Marine Biology

Berrill, N.J. and Jacquelyn. *1001 Questions Answered About the Seashore*. New York: Dodd, Mead & Company, 1957.
Hardy, Sir Alister. *The Open Sea: Its Natural History, Part I, The World of Plankton;* Part II, *Fish and Fisheries*. Boston: Houghton Mifflin Co., 1965.
Herald, Earl S. *Living Fishes of the World*. Garden City, N.Y.: Doubleday & Company, Inc., 1961.
Idyll, C. P. *Abyss—the Deep Sea and the Creatures That Live in It*. New York: Thomas Y. Crowell Co., 1964.
Kellogg, W. N. *Porpoises and Sonar*. Chicago: University of Chicago Press, 1966.
Matthews, Leonard H., ed. *The Whale*. New York: Simon and Schuster, 1968.
Riedman, Sarah R., and Gustafson, Elton T. *Focus on Sharks*. New York: Abelard-Schuman, 1969.
Scheffer, Victor B. *The Year of the Whale*. New York: Charles Scribner's Sons, 1969.
Silverberg, Robert. *The World of Coral*. New York: Duell, Sloan and Pearce, 1965.

Food From the Sea

Bardach, John E. *Harvest of the Sea*. New York: Harper & Row, 1968.
Hickling, C. F. *The Farming of Fish*. London: Pergamon Press, 1968.

Marine Resources

American Assembly, Columbia University. *Uses of the Sea*. Englewood Cliffs, N.J.: Prentice-Hall, Inc., 1968.

Mero, John L. *The Mineral Resources of the Sea*. Amsterdam: Elsevier Publishing Co., 1964.

Pollution

Krenkel, Peter A., and Parker, Frank L., eds. *Biological Aspects of Thermal Pollution*. Nashville, Tenn.: Vanderbilt University Press, 1969.

Myths and Legends

Cook, Joseph J., and Wisner, William L. *Warrior Whale*. New York: Dodd, Mead & Company, 1966.

Helm, Thomas. *Monsters of the Deep*. New York: Dodd, Mead & Company, 1962.

Knowlton, William. *Sea Monsters*. New York: Alfred A. Knopf, 1959.

For the Advanced Student

Riley, J. P., and Skirrow, G., eds. *Chemical Oceanography*. 2 vols. London: Academic Press, 1965.

Shepard, Francis P. *Submarine Geology*. New York: Harper & Row, 1963.

Sverdrup, H. U.; Johnson, Martin W.; and Fleming, Richard H. *The Oceans, Their Physics, Chemistry, and General Biology*. New York: Prentice-Hall, Inc., 1942.

Weyl, Peter K. *Oceanography: An Introduction to the Marine Environment*. New York: John Wiley & Sons, Inc., 1970.

Williams, Jerome. *Oceanography: An Introduction to the Marine Sciences*. Boston: Little, Brown and Co., 1962.

INDEX

References are to question numbers

abyssal plain, 94, 134
accelerometer, 258
Adriatic Sea, 292
Agassiz, Alexander, 42, 985
age of water, 316, 317
air embolism, 796
air-sea interaction, Chapter VIII, 402–19
Akron, University of, 630
Alaska, University of, 75, 77
Albatross, 42
Aleutian Trench, 136
alewives, 746
algae, 459, 481–83, 501, 611, 612, 912
Aluminaut, 830–32, 836
aluminum, 843
Alvin, 829–32, 836
ambergris, 588
ambient noise, 242
American Mediterranean, 19
anadromous fishes, 512, 746–48
anchor ice, 357
anchoring, 52
anchovies, 542, 689, 700, 708,735, 741, 755, 759
Andrea Doria, 812
angler fish, 517
aquaculture, 731, 884–95
aqualung, 798
Arabian Sea, 756
archeology, underwater, 816, 817
Archerfish, 139
Archimedes, 229, 826
arctic frost smoke, 416
Arctic Research Laboratory, 77
Aristotle, 287, 574, 975
ARLIS ice island, 77, 397, 399
arsenic, 841, 935
artificial seawater, 426
Atlantis, the lost continent, 972, 973

Atlantis, USS, 815
Atomic Energy Commission, 917, 943

Bacon, Sir Francis, 160
bacteria, 643, 644, 668, 913, 923
baleen whale, 557, 579, 760
Baltic Sea, 12, 292, 443, 445, 729
banks, 739
Barents Sea, 348, 701
barium, 841
barium spherules, 862
barnacles, 629, 630
barracuda, 554, 557, 568, 572
Barracuda, USS, 168
Barton, Otis, 823, 824
basins, ocean, 133
basking shark, 546, 977
Bass, George, 817
bass, striped, 727
Bathysphere, 823, 824
bathythermograph
    expendable electronic, 192, 193
    mechanical, 185, 190–92, 204
bay, 21
Bay of Fundy, 291
Beagle, 148
beam trawl, 764
Beebe, William, 540, 823, 824, 827
Bellingshausen Sea, 380
Bell Telephone Laboratories, 174
bench mark, 282
bends, 796
Ben Franklin, 833–35
benthos, 476–79
Benthosaurus, 525
bergy bits, 395
Bible, 27, 903, 994, 1001
Bigelow, Henry B., 73
biological oxygen demand, 957
biology, marine, Chapter X, 457–652